Author: Brenda Gunderson, Ph.D., 2014

Mind on Statistics
Utts/Heckard, 4th Edition, Cengage L, 2012
Text Only: ISBN 9781285135984
Bundled version: ISBN 9780538733489

Statistics 250

Interactive Lecture Notes

Fall 2014 – Winter 2015

Dr. Brenda Gunderson
Department of Statistics
University of Michigan

Michigan Publishing
University of Michigan Library
Ann Arbor

ISBN 978-1-60785-438-8

Table of Contents

Stat 250 Gunderson Lecture Notes
Chapter 1: Statistics Success Stories

Statistics...**the most important science in the whole world**: for upon it depends the practical application of every other science and of every art: the one science essential to all political and social administration, all education, all organization based on experience, for it only gives results of our experience." *Florence Nightingale, Statistician* © FAIR USE

> ***Definitions:***
> **Statistics** are numbers measured for some purpose.
> **Statistics** is a collection of procedures and principles for gathering data and analyzing information in order to help people make decisions when faced with uncertainty.

Course Goal: Learn various tools for using data to gain understanding and make sound decisions about the world around us.

Chapter 1 starts out with eight statistical stories with morals, presented as seven case studies. In each, data are used to make a decision, a judgment, about a situation. These case studies follow a wide range of ideas and methods and introduce a lot of statistical language. Take a few minutes to read them and add your own summary regarding the moral to each story below.

Case Study	Moral
1: Who are those speedy drivers?	
2: Safety in the skies?	
3: Did anyone ask whom you've been dating?	
4: Who are those angry women?	
5: Does prayer lower blood pressure?	
6: Does aspirin reduce heart attack rates?	
7: Does the internet increase loneliness and depression?	
8: Did your mother's breakfast determine your sex?	

Additional Notes

A place to ... jot down questions you may have and ask during office hours, take a few extra notes, write out an extra practice problem or summary completed in lecture, create your own short summary about this chapter.

Stat 250 Gunderson Lecture Notes
Chapter 2: Turning Data into Information

" You must never tell a thing. You must illustrate it. We learn through the eye and not the noggin."

--Will Rogers (1879 - 1935) © FAIR USE

Chapter 2 focuses on the moral of Case Study 1, namely, **"Simple summaries of data can tell an interesting story and are easier to digest than long lists."** So we will begin by looking at some data.

2.1 Raw Data

Raw data correspond to numbers and category labels that have been collected or measured but have not yet been *processed* in any way. On the next page is a set of RAW DATA - information about a group of items in this case, individuals. The data set title is DEPRIVED and has information about a **sample size** of ***n*** = 86 college students. For each student we are provided with their answer to the question: "Do you feel that you are sleep deprived?" (yes or no), and their self reported typical amount of sleep per night (in hours). The information we have is organized into **variables**. In this case these 86 college students are a subset from a larger population of all college students, so we have **sample data**.

Definition:
A **variable** is a characteristic that differs from one individual to the next.
Sample data are collected from a subset of a larger population.
Population data are collected when all individuals in a population are measured.
A **statistic** is a summary measure of sample data.
A **parameter** is a summary measure of population data.

2.2 Types of Variables

We have 2 variables in our data set. Next we want to distinguish between the different types of variables - different types of variables provide different kinds of information and the type will guide what kinds of summaries (graphs/numerical) are appropriate.

Think about it:

- Could you compute the "AVERAGE AMOUNT OF SLEEP" for these 86 students?
- Could you compute the "AVERAGE SLEEP DEPRIVED STATUS" for these 86 students?

SLEEP DEPRIVED STATUS is said to be a ________________________ variable,

AMOUNT OF SLEEP is a ____________________ variable.

> ***Definitions:***
> A **categorical** variable places an individual or item into one of several groups or categories. When the categories have an ordering or ranking, it is called an **ordinal** variable.
> A **quantitative** variable takes numerical values for which arithmetic operations such as adding and averaging make sense. Other names for quantitative variable are: **measurement** variable and **numerical** variable.

Try It! –

For each variable listed below, give its type as categorical or quantitative.

- Age (years)
- Typical Classroom Seat Location (Front, Middle, Back)
- Number of songs on an iPod
- Time spent studying material for this class in the last 24-hour period (in hours)
- Soft Drink Size (small, medium, large, super-sized)
- The "And then ..." count recorded in a psychology study on children (details will be provided)

> ***Looking ahead:*** Later, when we talk about random variables, we will discuss whether a variable is modeled *discretely* (because it's values are countable) or whether it would be modeled *continuously* (because it can take any value in an interval or collection of intervals). Go back through the list above and think about is it *discrete* or *continuous*?

DATA SET = DEPRIVED

Source: Utts MOS, Deprived Data Set

Feel Sleep Deprived?	Amount Sleep per Night (hours)
No	8.0
No	6.0
No	7.0
Yes	6.0
Yes	6.0
Yes	7.0
Yes	6.0
Yes	7.0
No	9.0
No	7.0
No	8.0
No	7.0
Yes	7.0
Yes	3.0
Yes	5.0
No	7.0
No	9.0
No	3.0
Yes	6.0
Yes	7.0
No	8.0
No	8.0
No	6.0
Yes	7.0
No	8.0
No	8.0
No	7.0
No	5.0
No	8.0
Yes	6.0
Yes	10.0
Yes	6.0
No	8.0
Yes	6.0
No	7.0
Yes	6.0
Yes	6.0
Yes	8.0
Yes	0.0
Yes	6.0
Yes	5.0
No	7.0
Yes	5.0

Feel Sleep Deprived?	Amount Sleep per Night (hours)
No	7.0
No	6.0
No	8.0
Yes	6.0
Yes	6.0
Yes	6.0
Yes	6.0
Yes	5.0
No	7.0
Yes	5.0
No	8.0
No	7.0
Yes	6.0
Yes	7.0
No	7.0
No	7.0
Yes	6.0
Yes	6.0
Yes	6.0
No	6.0
Yes	6.0
Yes	7.0
No	6.0
Yes	6.0
Yes	6.0
Yes	6.0
Yes	7.0
Yes	5.0
Yes	5.0
Yes	7.0
No	8.0
No	6.0
Yes	7.0
Yes	5.0
Yes	6.0
Yes	7.0
Yes	4.0
Yes	5.0
No	6.0
No	7.0
Yes	7.0
Yes	6.0
Yes	5.0

Our data set is somewhat large, containing a lot of measurements in a long list. Presented as a table listing, we can view the record of a particular college student, but it is just a listing, and not easy to find the largest value for the amount of sleep or the number of students who felt they are sleep deprived. We would like to learn appropriate ways to summarize the data.

2.3 Summarizing One or Two Categorical Variables

Numerical Summaries

How would you go about summarizing the SLEEP DEPRIVED STATUS data? The first step is to simply count how many individuals/items fall into each category. Since percents are generally more meaningful than counts, the second step is to calculate the percent (or proportion) of individuals/items that fall into each category.

Sleep Deprived?	Count	Percent
Yes		
No		
Total		

The table above provides both the **frequency distribution** and the **relative frequency distribution** for the variable SLEEP DEPRIVED STATUS.

Visual Summaries

There are two simple visual summaries for categorical data – a bar graph or a pie chart. Here is the table summary and bar graph made with SPSS (Analysis → Frequencies).

Sleep Deprived?

		Frequency	Percent
Valid	No	35	40.7
	Yes	51	59.3
	Total	86	100.0

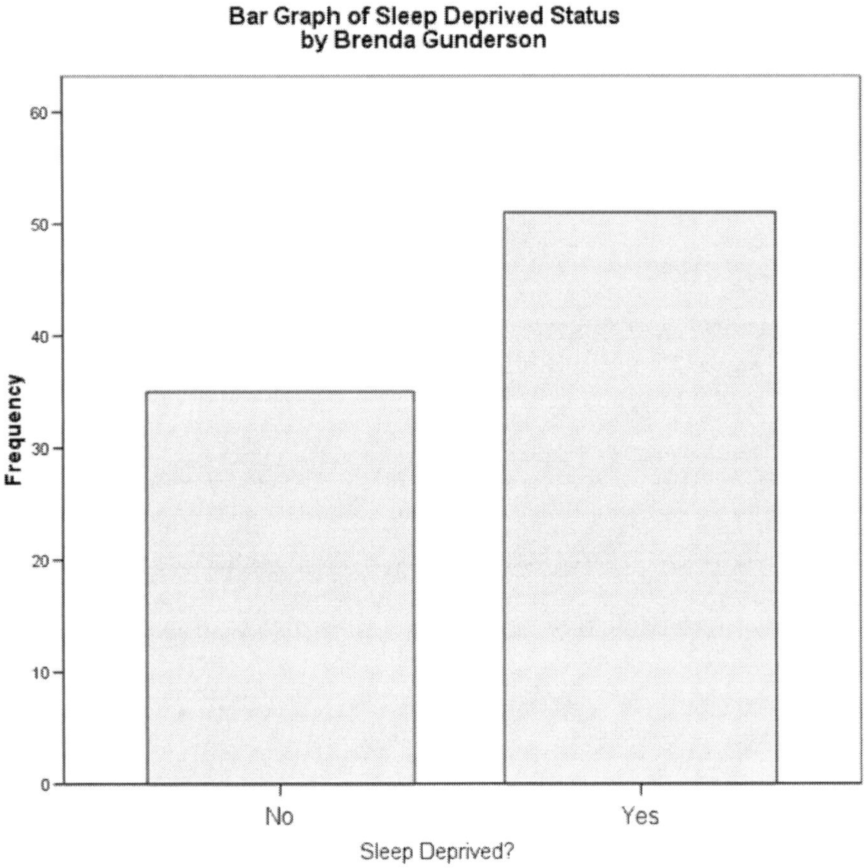

Aside: Does it matter whether the 'No' or 'Yes' bar is given first?

Pie Chart: Another graph for categorical data which helps us see what part of the whole each group forms.

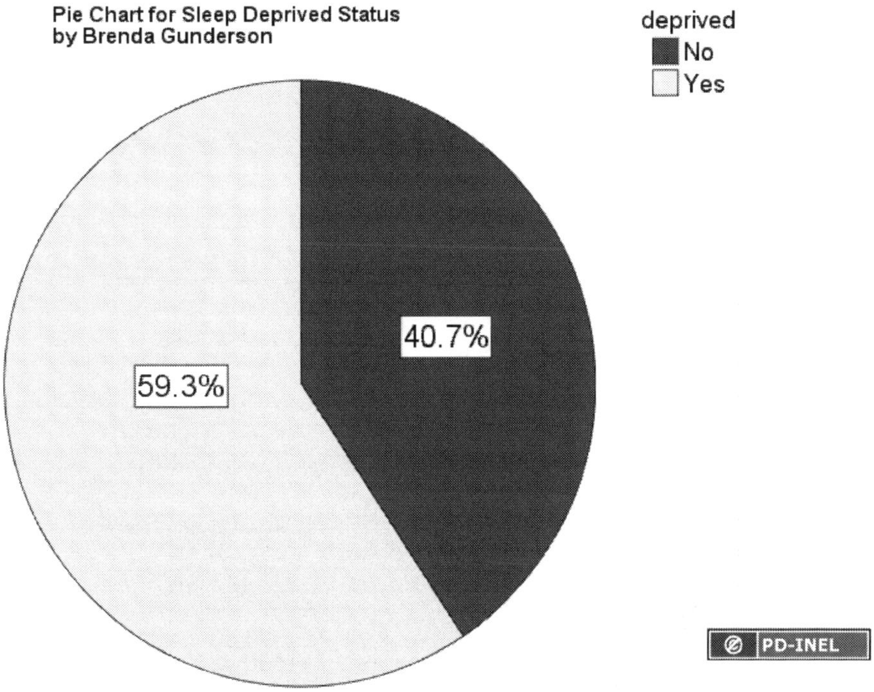

Pie charts are not as easy to draw by hand. It is not as easy to compare sizes of pie pieces versus comparing heights of bars.

Thus we will prefer to use a bar graph for categorical data.

Recap: We have discussed that some variables are categorical and others are quantitative. We have seen that bar graphs and pie charts can be used to display data for categorical variables. We turn next to displaying the data for **quantitative** variables.

2.4 Exploring Features of Quantitative Data with Pictures

Recall our Sleep Deprived Data for n = 86 college students. We have data on two variables: sleep deprived status and hours of sleep per night. How would you go about summarizing the sleep hours data? These measurements do vary. How do they vary? What is the range of values? What is the pattern of variation?

- Find the smallest value = ____________ and largest value = ____________
- Take this overall range and break it up into intervals (of equal width). What might be reasonable here? Perhaps by 2's; but we need to watch the endpoints.

Summary Table:

Class	Frequency (or count)	Relative Frequency (or proportion)	Percent
0, 2			
2, 4			
4, 6			
6, 8			
8, 10			

Graph for quantitative data = Histogram:

Note: each bar represents a class, and the base of the bar covers the class.
The above table and histogram show the **<u>distribution</u> of this quantitative variable** *SLEEP HOURS*, that is, the overall pattern of how often the possible values occur.

SPSS Histograms (default on top and customized on bottom):

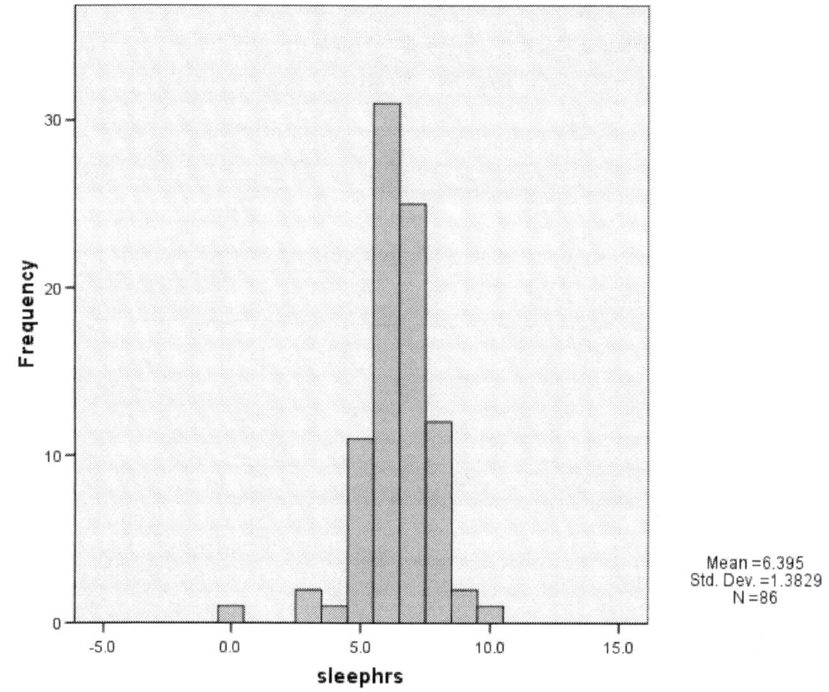

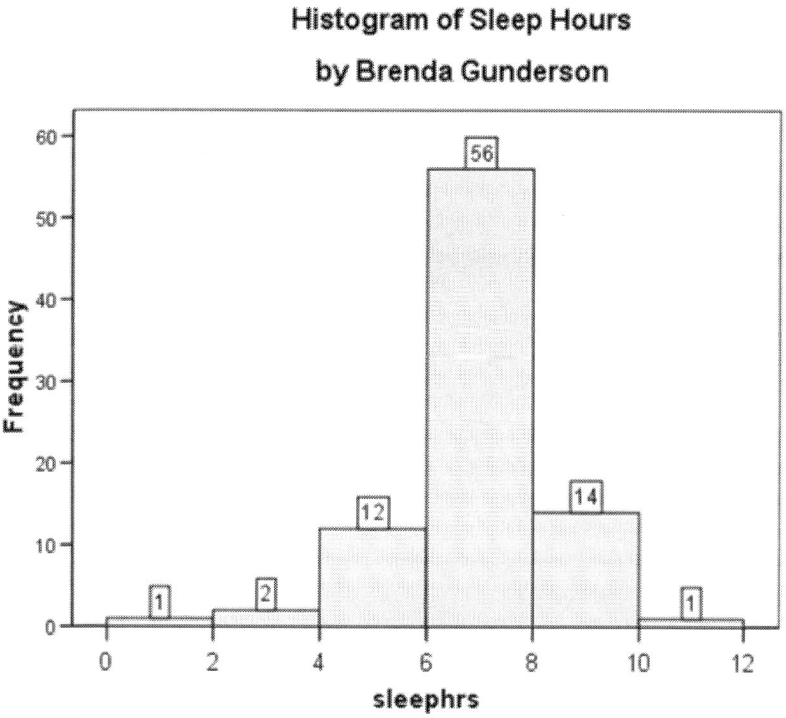

How to interpret?

- **Look for Overall Pattern**
  Three summary characteristics of the overall distribution of the data ...
  Shape (approximately symmetric, skewed, bell-shaped, uniform)

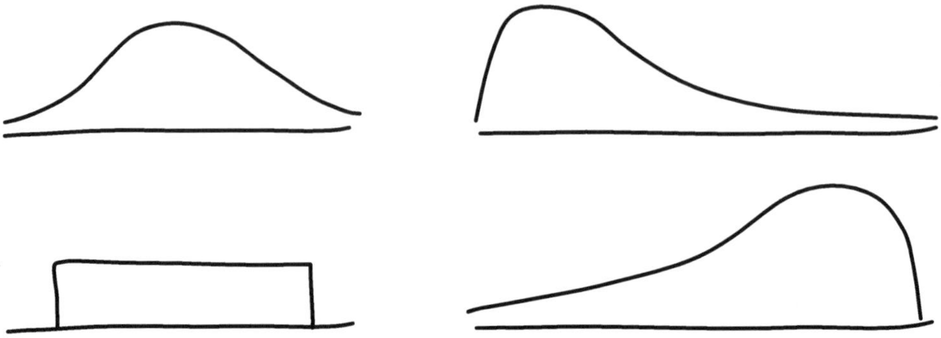

  Location (center, average)

  Spread (variability)

- **Look for deviations from Overall Pattern**
  Outliers = a data point that is not consistent with the bulk of the data. Outliers should not be discarded without justification (see Section 2.6 of the text).

Describe the distribution for SLEEP HOURS:

What if ...
you had some data and you made a histogram of it and it looked like this...

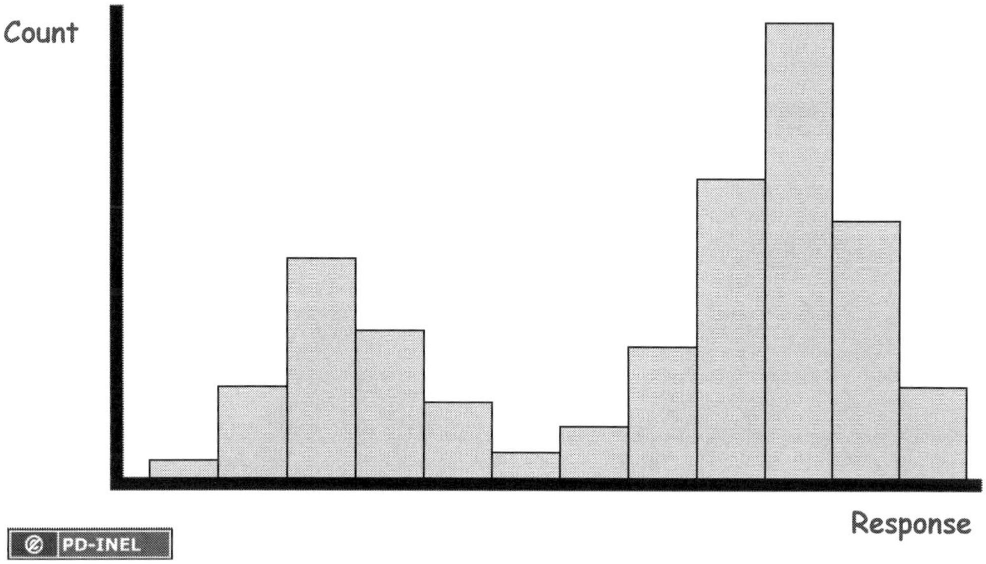

What would it tell you?

Other comments –

- NO SPACE BETWEEN BARS! Unless there are no observations in that interval.
- How Many Classes? Use your judgment: generally somewhere between 6 and 15 intervals.
- Better to use relative frequencies on the y axis when comparing two or more sets of observations.
- Software has defaults and many options to modify settings.

Textbook Notes:

We will not cover the making of dotplots or stem-and-leaf plots but you should read the entire Sections of 2.4 and 2.5. All three graphs would provide a similar picture of the distribution and are useful for assessing center, spread, shape, and outliers. Boxplots are also briefly introduced in Section 2.4 but covered in more detail in Section 2.5. Pages 35-36 provide a good summary of the various visual displays for quantitative data.

One More Example:

A study was conducted in Detroit, Michigan to find out the number of hours children aged 8 to 12 years spent watching television on a typical day.

A listing of all households in a certain housing area having children aged 8 to 12 years was first constructed. Out of the 100 households in this listing, a random sample of 20 households was selected and all children aged 8 to 12 years in the selected households were interviewed.

The following histogram was obtained for all the children aged 8 to 12 years interviewed.

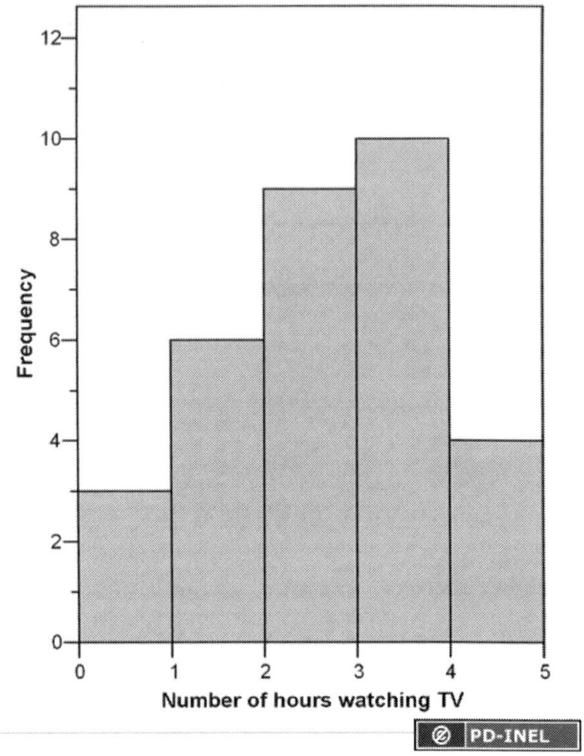

a. Complete the sentence: Based on this histogram, the distribution of number of hours spent watching TV is unimodal,

   with a slight skewness to the __________.

b. Assuming that all children interviewed are represented in the histogram, what is the total number of children interviewed?

c. What proportion of children spent less than 2 hours watching television?

d. Can you determine the maximum number of hours spent watching television by one of the interviewed children? If so, report it. If not, explain why not.

2.5 Numerical Summaries of Quantitative Variables

We have discussed some interesting features of a quantitative data set and learned how to look for them in pictures (graphs). Section 2.5 focuses on numerical summaries of the center and the spread of the distribution (appropriate for quantitative data only).

Notation for a generic raw set of data:

$x_1, x_2, x_3, \ldots, x_n$ **where n = # items in the data set or sample size**

Describing the Location or Center of a Data Set

Two basic measures of location or center:

- **Mean** -- the numerical average value
  We represent the ***mean of a sample*** (called a statistic) by ...

$$\bar{x} = \frac{x_1 + x_2 + \cdots + x_n}{n} = \frac{\sum x_i}{n}$$

- **Median** -- the middle value when data arranged from smallest to largest.

Fries Example:

Weight measurements for 12 small orders of French fries (in grams).

77	72	69	80	63	67
78	74	70	83	71	79

What should we do with data first? Graph it!

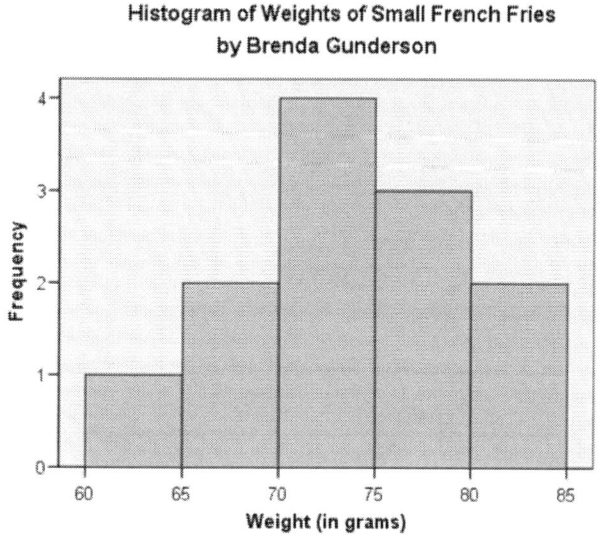

Based on our histogram, the distribution of weight is unimodal and approximately symmetric, so computing numerical summaries is reasonable. The weights (in grams) range from the 60's to the lower 80's, centered around the lower 70's.

1. **Compute the mean weight.**

2. **Compute the median weight.**
   Ordered: 63, 67, 69, 70, 71, 72, 74, 77, 78, 79, 80, 83

3. **What if the smallest weight was incorrectly entered as 3 grams instead of 63 grams?**

Note: The **mean** is ____________________ to extreme observations.

The **median** is ________________________ to extreme observations.

Most graphical displays would have detected such an outlying value.

Some Pictures: Mean versus Median

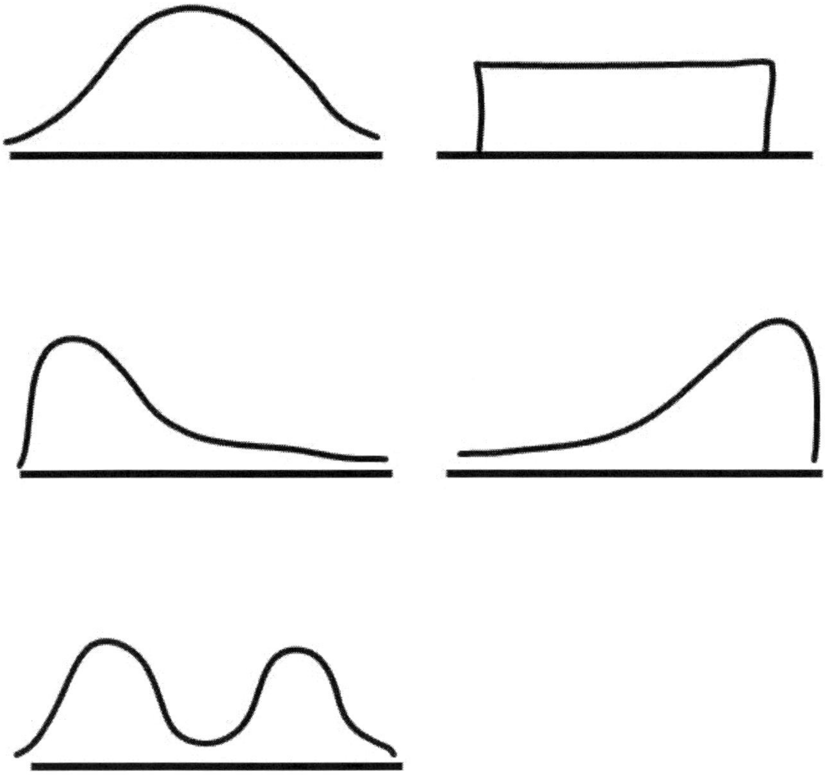

Describing Spread: Range and Interquartile Range

Midterms are returned and the "average" was reported as 76 out of 100.
You received a score of 88.
How should you feel?

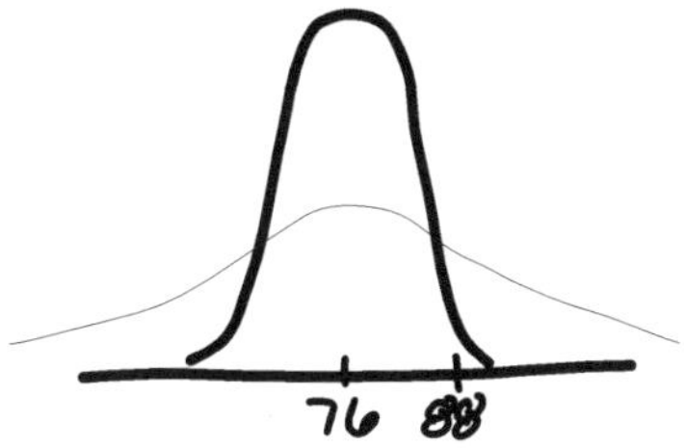

Often what is missing when the "average" of something is reported, is a corresponding measure of spread or variability. Here we discuss various measures of variation, each useful in some situations, each with some limitations.

Range: Measures the spread over 100% of the data.
Range = High value – Low value = Maximum – Minimum

Percentiles: The pth percentile is the value such that p% of the observations fall at or below that value.

Some Common percentiles:
Median: 50th percentile

First quartile: 25th percentile

Third quartile: 75th percentile

Five Number Summary:

	Variable Name and Units (n = number of observations)		
Median		M	
Quartiles	Q1		Q3
Extremes	Min		Max

Provides a quick overview of the data values and information about the center and spread. Divides the data set into approximate quarters.

Interquartile Range: Measures the spread over the middle 50% of the data.
IQR = Q3 – Q1

Try it! French Fries Data

Ordered: 63, 67, 69, 70, 71, 72, 74, 77, 78, 79, 80, 83

Find the five-number summary:

	Weight of Fries (in grams) (n = 12 orders)		
Median			
Quartiles			
Extremes			

Range:

IQR:

Example: Test Scores

The five-number summary for the distribution of test scores for a very large math class is provided below:

	Test Score (points) (n = 1200 students)		
Median		58	
Quartiles	46		78
Extremes	34		95

1. What is the test score interval containing the lowest ¼ of the students?

2. Suppose you scored a 46 on the test. What can you say about the percentage of students who scored higher than you?

3. Suppose you scored a 50 on the test. What can you say about the percentage of students who scored higher than you?

4. If the top 25% of the students received an A on the test, what was the minimum score needed to get an A on the test?

Boxplots

A boxplot is a graphical representation of the five-number summary.

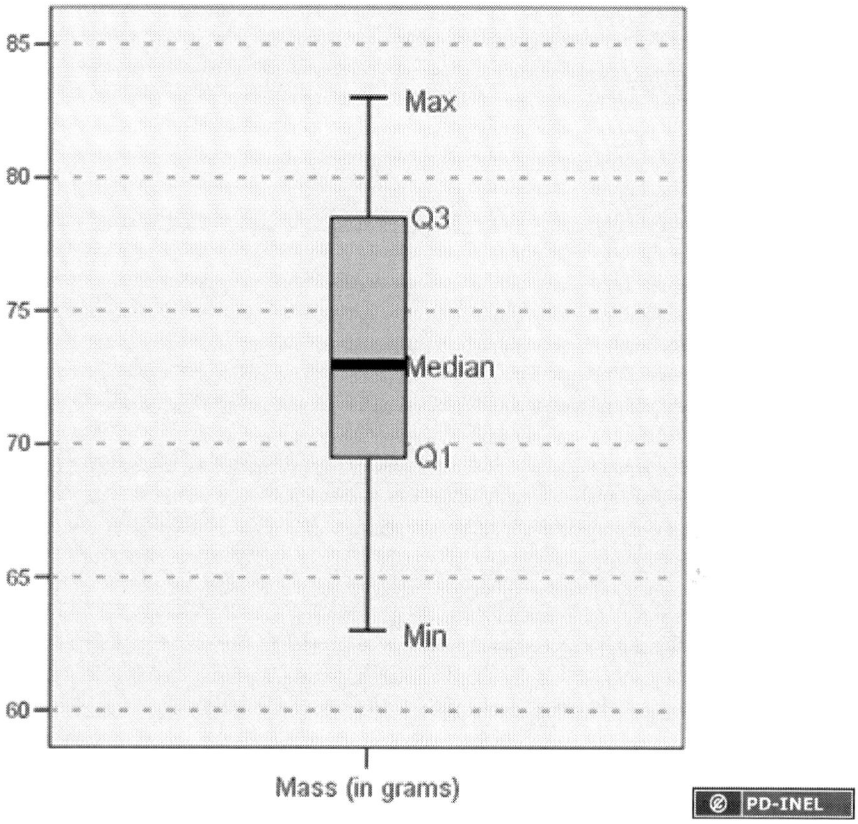

Steps:

- **Label an axis with values to cover the minimum and maximum of the data.**
- **Make a box with ends at the quartiles Q1 and Q3.**
- **Draw a line in the box at the median *M*.**
- **Check for possible outliers using the 1.5*IQR rule and if any, plot them individually.**
- **Extend lines from end of box to smallest and largest observations that are not possible outliers.**

Note: Possible outliers are observations that are more than 1.5*IQR outside the quartiles. That is, observations that are below Q1 - 1.5*IQR or observations that are above Q3 + 1.5*IQR.

Try it! French Fries Data

Ordered data: 63, 67, 69, 70, 71, 72, 74, 77, 78, 79, 80, 83

The five-number summary:

	Weight of Fries (in grams) (*n* = 12 orders)		
Median		73	
Quartiles	69.5		78.5
Extremes	63		83

From the boxplot shown, we see there are no points plotted separately, so there are no outliers by the 1.5(IQR) rule.

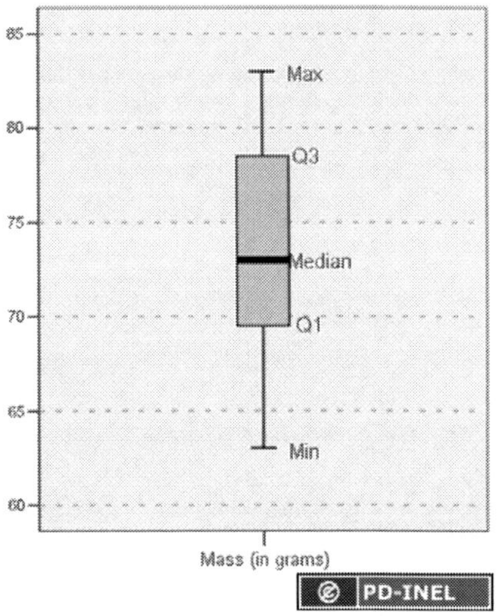

Verify there are no outliers using this rule.

IQR = 78.5 – 69.5 = 9 grams

1.5*IQR = 1.5 (9) = 13.5 grams

Lower boundary (fence) = Q1 - 1.5*IQR =

Are there any observations that fall below this lower boundary?

Upper boundary (fence) = Q3 + 1.5*IQR =

Are there any observations that fall above this upper boundary?

What if ... the largest weight of 83 grams was actually 93 grams?

Ordered: **63, 67, 69, 70, 71, 72, 74, 77, 78, 79, 80, *93***

Then the five number summary would be:

	Weight of Fries (in grams) (*n* = 12 orders)		
Median		73	
Quartiles	69.5		78.5
Extremes	63		

The IQR and 1.5*IQR would be the same, so the "boundaries" for checking for possible outliers are again 56 and 92.

Now we would have one potential high outlier, the maximum value of 93.

The modified boxplot when we have this one outlier is shown below.
Why is the line extending out on the top side now drawn out to just 80?

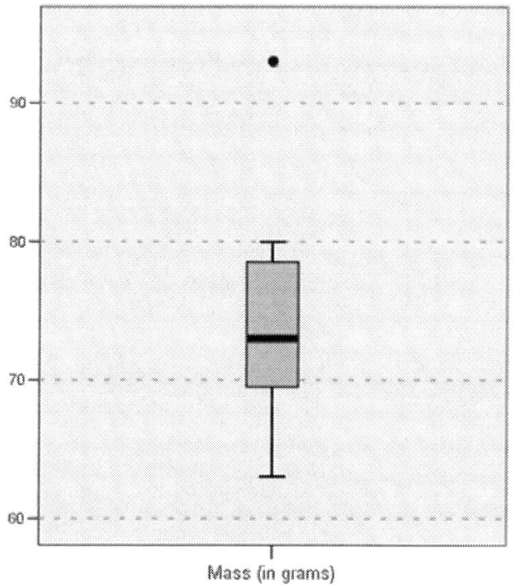

Notes on Boxplots:

- **Side-by-side boxplots are good for ...**

- **Watch out - points plotted individually are ...**

- **Can't confirm ....**

- **When reading values from a graph show what you are doing (so appropriate credit can be given on exam/quiz).**

Try It: Side-by-side boxplots

A random sample of 100 parents of grade-school children were recently interviewed regarding the breakfast habits in their family. One question asked was if their children take the time to eat a breakfast (recorded as breakfast status – Yes or No). The grades of the children in some core classes (e.g. reading, writing, math) were also recorded and a standardized grade score (on a 10-point scale) was computed for each child. Side-by-side boxplots of the children's standardized grade scores are provided.

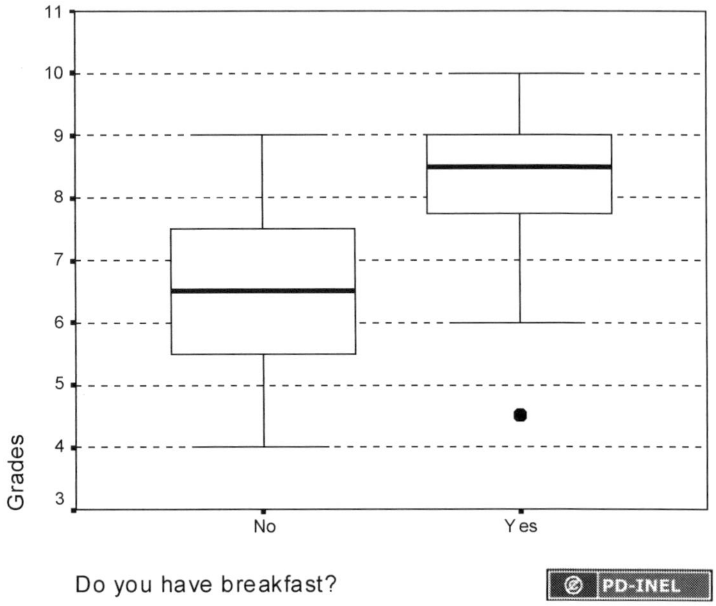

a. What is (approx) the lowest grade scored by a child who **does** have breakfast?

   ___________ points

b. Complete the following sentence:

   Among the children who did **not** eat breakfast,
   25% had a grade score of ***at least*** ______________ points.

c. Consider the following statement: The symmetry in the boxplot for the children **not** eating breakfast *implies* that the distribution for the grade scores of such students is bell-shaped.

   True or False?

2.6 How to Handle Outliers

IN SUMMARY Possible Reasons for Outliers and Reasonable Actions

- *The outlier is a legitimate data value and represents natural variability for the group and variable(s) measured.* Values may not be discarded in this case — they provide important information about location and spread.
- *A mistake was made while taking a measurement or entering it into the computer.* If this can be verified, the values should be discarded or corrected.
- *The individual in question belongs to a different group than the bulk of individuals measured.* Values may be discarded if a summary is desired and reported for the majority group only.

From Utts, Jessica M. and Robert F. Heckard. Mind on Statistics, Fourth Edition. 2012. Used with permission.

2.7 Features of Bell-Shaped Distributions

We have already described the distribution of our french fries weight data as being unimodal and somewhat symmetric. If we were to draw a curve to smooth out the tops of the bars of the histogram, it would resemble the shape of a bell, and thus could be called **bell-shaped**.

One fairly common distribution of measurements with this shape has a special name, called a **normal distribution** or **normal curve**. We will see normal curves in more detail in Chapter 8. When a distribution is somewhat bell-shaped (unimodal, indicating a fairly homogeneous set of measurements), a useful measure of spread is called the **standard deviation**. In fact, the mean and the standard deviation are two summary measures that completely specify a normal curve.

Describing Spread with Standard Deviation

When the mean is used to measure center, the most common measure of spread is the standard deviation. The standard deviation is a ***measure of the spread of the observations from the mean***. We will refer to it as a kind of "average distance" of the observations from the mean. But it actually is the square root of the average of the squared deviations of the observations from the mean. Since that is a bit cumbersome, we like to **think of the standard deviation as "roughly, the average distance the observations fall from the mean."** Here is a quick look at the formula for the stand deviation when the data are a sample from a larger population:

***s* = sample standard deviation =**

Note: The squared standard deviation, denoted by s^2, is called the **variance**. We emphasize the standard deviation since it is in the original units.

Example: Consider this sample of $n = 5$ scores: 94, 97, 99, 103, 107.
The sample mean is 100 points. Let's measure their *spread* by considering how much each score ***deviates* from the mean**. Then consider the "average of these deviations".

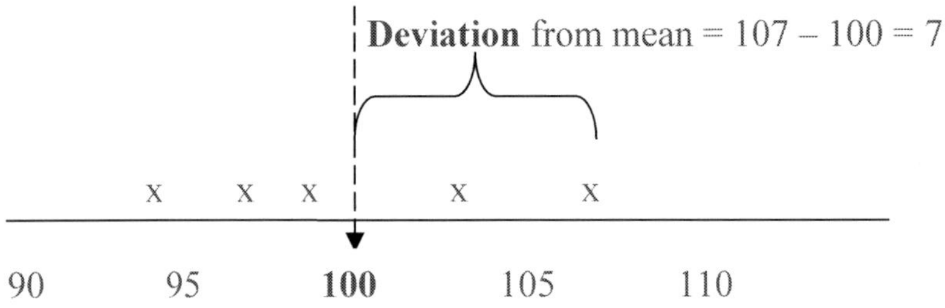

How the calculations are done:

x	x – 100	$(x-100)^2$	Calculations
94	-6	36	1. $\bar{x} = 500/5 = 100$ (used in columns 2 & 3)
97	-3	9	2. Variance: $s^2 = 104/(5-1) = 26$
99	-1	1	**3. Standard deviation: $s = \sqrt{(26)} = 5.1$**
103	3	9	Note # of *deserved* decimals used for s.
107	7	49	
500	**0**	**104**	← Sums (or totals) of the columns

Fries Example:
Weight measurements for 12 small orders of French fries (in grams).

77 72 69 80 63 67 78 74 70 83 71 79

The mean was computed earlier to be 73.6. **Find the standard deviation** for this data.

$s =$

Not much fun to do it by hand, but not too bad for a small number of observations. In general, we will have a calculator or computer do it for us.

Interpretation:

The weights of small orders of french fries are *roughly*

_______________away from their mean weight of _______, *on average*.

OR

On average, the weights of small orders of french fries vary by about ________ from their mean weight of ______________________.

Notes about the standard deviation:

- ***s*** **= 0 means ...**

- **Like the mean,** ***s*** **is ...**

- **So use the mean and standard deviation for__________________________________.**

  The five-number summary is better for ____.

- **Technical Note about difference between population and sample:** Datasets are commonly treated as if they represent a sample from a larger population. A numerical summary based on a sample is called a statistic. The sample mean and sample standard deviation are two such statistics. However, if you have all measurements for an entire population, then a numerical summary would be referred to as a parameter.

  The symbols for the mean and standard deviation for a population are different, and the formula for the standard deviation is also slightly different. A population mean is represented by the Greek letter μ ("mu"), and a population standard deviation is represented by the Greek letter σ ("sigma"). The formula for the population standard deviation is below.

  You will see more about the distinction between statistics and parameters in the next chapter and beyond.

  Population standard deviation: $\sigma = \sqrt{\dfrac{\sum (x_i - \mu)^2}{N}}$

  where ***N*** **is the size of the population.**

Empirical Rule

For bell-shaped curves, approximately...

68% of the values fall within 1 standard deviation of the mean in either direction.

95% of the values fall within 2 standard deviations of the mean in either direction.

99.7% of the values fall within 3 standard deviations of the mean in either direction.

Exercises 2.85 and 2.89 page 62-63*

From Utts, Jessica M. and Robert F. Heckard. Mind on Statistics, Fourth Edition. 2012. Used with permission.

The typical amount of sleep per night for college students has a bell-shaped distribution with a mean of 7 hours and a standard deviation of 1.7 hours.

About 68% of college students typically sleep between ________ and ________ hours per night.

Verify the values below that complete the sentences.

- About 95% of college students typically sleep between **3.6** and **10.4** hours per night.
- About 99.7% of college students typically sleep between **1.9** and **12.1** hours per night.

Draw a picture of the distribution showing the mean and intervals based on the empirical rule.

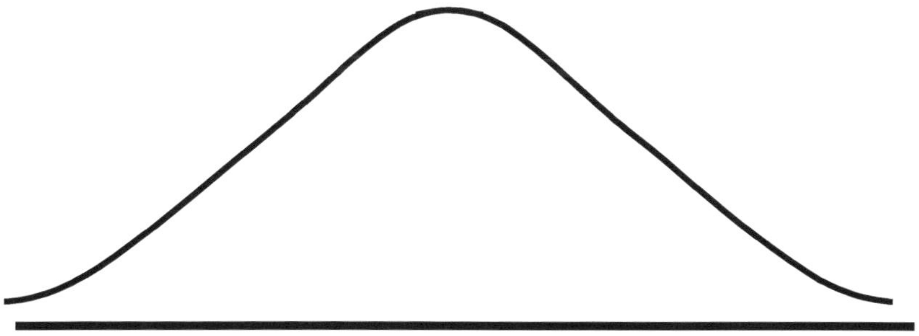

Suppose last night you slept 11 hours.
How many standard deviations from the mean are you?

Suppose last night you slept only 5 hours.
How many standard deviations from the mean are you?

The standard deviation is a useful "**yardstick**" for measuring how far an individual value falls from the mean. The **standardized score** or ***z*-score** is the distance between the observed value and the mean, measured in terms of number of standard deviations. Values that are above the mean have positive *z*-scores, and values that are below the mean have negative *z*-scores.

Standardized score or *z*-score: $z = \frac{\text{observed value} - \text{mean}}{\text{standard deviation}}$

Empirical Rule (in terms of z-scores)

For bell-shaped curves, approximately...

68% of the values have *z*-scores between –1 and 1.
95% of the values have *z*-scores between –2 and 2.
99.7% of the values have *z*-scores between –3 and 3.

Example: Scores on a Final Exam

Scores on the final in a course have approximately a bell-shaped distribution. The mean score was 70 points and the standard deviation was 10 points.

Suppose Rob, one of the students, had a score that was 2 standard deviations above the mean. What was Rob's score?

What can you say about the proportion of students who scored higher than Rob?

Sketching a picture may help.

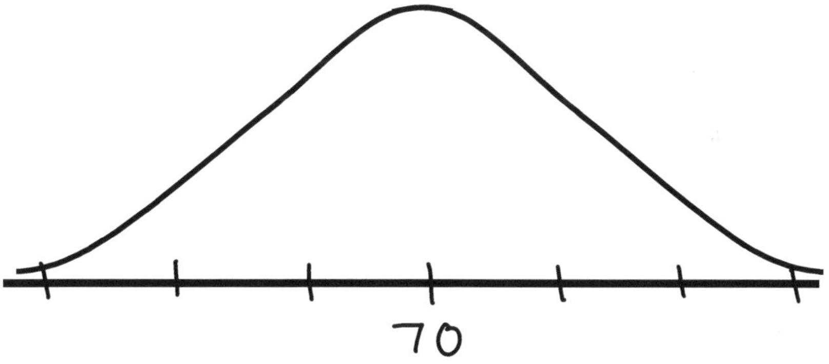

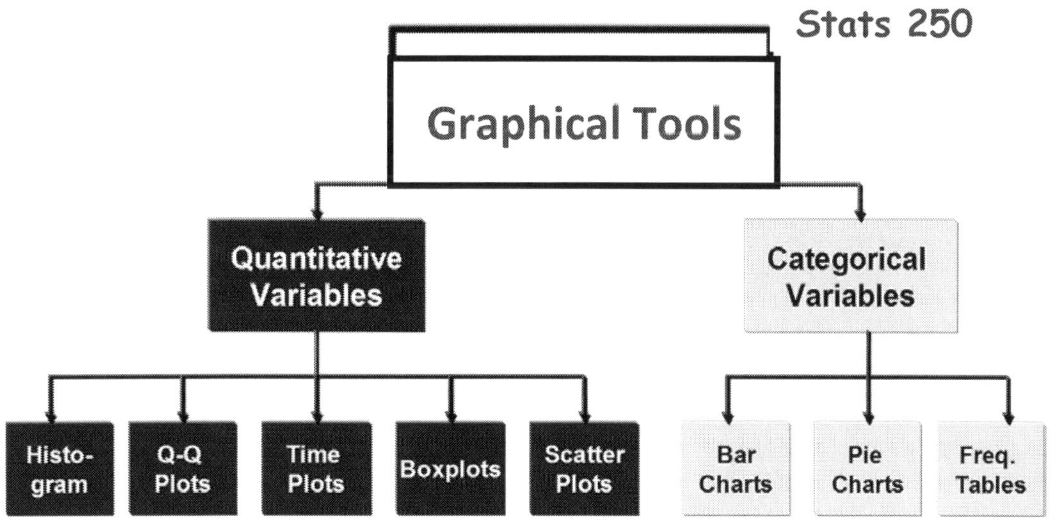

Additional Notes

A place to … jot down questions you may have and ask during office hours, take a few extra notes, write out an extra practice problem or summary completed in lecture, create your own short summary about this chapter.

Stat 250 Gunderson Lecture Notes
Chapter 5: Sampling: Surveys and How to Ask Questions

Do not put faith in what statistics say until you have carefully considered what they do not say. -- *William W. Watt*

So far we have mainly studied how to summarize data - exploratory data analysis - with graphs and numbers. The knowledge of ***how the data were generated*** is one of the key ingredients for translating data intelligently. Chapters 5 and 6 focus on how to collect useful data. Chapter 5 focuses on how to conduct surveys, how to make sure they are representative, and what can go wrong. We will highlight some ideas here. We will not go in to detail on the various sampling methods in Sections 5.3 and 5.4, but will comment on random sampling. The sampling ideas in Chapter 5 overall make sense and there are a number of good examples. And there will be a few questions from this chapter on the homework and exams.

5.1 Collecting and Using Sample Data Wisely

There are two main types of statistical techniques that can be applied to data.

Definitions:

Descriptive Statistics: Describing data using numerical summaries (such as the mean, IQR, etc.) and graphical summaries (such as histograms, bar charts, etc.).

Inferential Statistics: Using sample information to make conclusions about a larger group of items/individuals than just those in the sample.

In most statistical studies, the objective is to use a small group of units (the sample) to make an inference (a decision or judgment) about a larger group (the population).

Definitions:
Population: The entire group of items/individuals that we want information about, about which inferences are to be made.
Sample: The smaller group, the part of the population we actually examine in order to gather information.
Variable: The characteristic of the items or individuals that we want to learn about.

From Utts, Jessica M. and Robert F. Heckard. Mind on Statistics, Fourth Edition. 2012. Used with permission.

One way to view these terms is through a *Basket Model*:

Population= basket of balls, 1 ball for each unit in population.

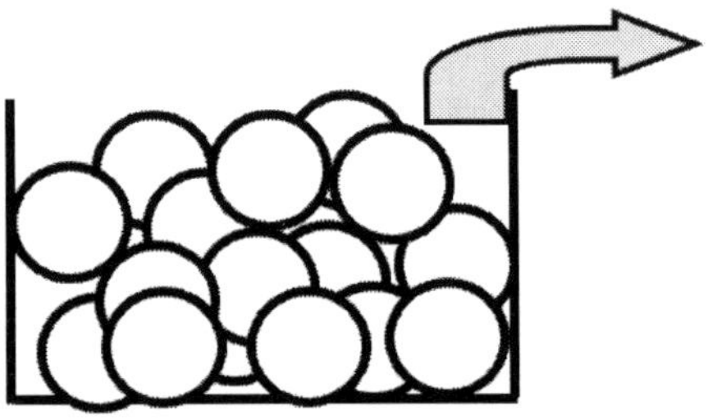

Sample = a few balls selected from the basket.

X **= variable (value of variable is recorded on each ball as small *x*)**

Fundamental Rule for Using Data for Inference:
Available data can be used to make inferences about a much larger group *if the data can be considered to be representative with regard to the question(s) of interest.*

One principal way to guarantee that sample data represents a larger population is to use a (simple) random sample.

Try It! Exercise 5.18 and 5.19 page 177

For each situation explain whether or not the Fundamental Rule holds.

From Utts, Jessica M. and Robert F. Heckard. Mind on Statistics, Fourth Edition. 2012. Used with permission.

1. **Research Question:** Does a majority of adults in state support lowering the drinking age to 19?
   Available Data: Opinions on whether or not the legal drinking age should be lowered to 19 years old, collected from a random sample of 1000 adults in the state.

2. **Research Question:** Does a majority of adults in state support lowering the drinking age to 19?
   Available Data: Opinions on whether or not the legal drinking age should be lowered to 19 years old, collected from a random sample of parents of high school students in the state.

3. **Available Data:** Pulse rates for smokers and nonsmokers in a large stats class at a major university.
   Research Question: Do college-age smokers have higher pulse rates than college-age nonsmokers?

Sample versus Census

Why can't we learn about a population by just taking a **census** (measure every item in the population)? Takes too long, costs too much, measuring destroys the item. So, we often rely on a special type of statistical study called a **sample survey**, in which a subgroup of a large population is questioned on a set of topics.

Sample surveys are often used to estimate the proportion or percentage of people who have a certain trait or opinion. If you use proper methods to sample 1500 people from a population of many millions, you can almost certainly gauge the percentage of the entire population who have a certain trait or opinion to within 3%. The tricky part is that you have to use a proper sampling method.

Bias: How Surveys Can Go Wrong

While it is unlikely that the sample value will equal the population value exactly, we do want our surveys to be unbiased. Results based on a survey are **biased** if the method used to obtain those results would consistently produce values that are either too high or too low.

Selection bias occurs if the method for selecting the participants produces a sample that does not represent the population of interest.

Nonparticipation bias (nonresponse bias) occurs when a representative sample is chosen for a survey, but a subset cannot be contacted or does not respond.

Biased response or response bias occurs when participants respond differently from how they truly feel. The way questions are worded, the way the interviewer behaves, as well as many other factors might lead an individual to provide false information.

From Utts, Jessica M. and Robert F. Heckard. Mind on Statistics, Fourth Edition. 2012. Used with permission.

Try It! Exercise 5.16 (page 177)

Which type of bias **do you think would be introduced if ...**

d. A magazine sends a survey to a random sample of its subscribers asking them if they would like the frequency of publication reduced from biweekly to monthly, or would prefer that it remain the same.

e. A random sample of registered voters is contacted by phone and asked whether or not they are going to vote in the upcoming presidential election.

5.2 Margin of Error, Confidence Intervals, and Sample Size

Sample surveys are often used to estimate the proportion or percentage of all people who have a certain trait or opinion (p). Newspapers and magazines routinely survey only one or two thousand people to determine public opinion on current topics of interest.

When a survey is used to find a proportion based on a sample ($\hat{p}$) of only a few thousand individuals, one question is ***how close that proportion comes to the truth for the entire population***. This measure of accuracy in sample surveys is a number called the **margin of error**.

The margin of error provides an upper limit on the amount by which the sample proportion $\hat{p}$ is expected to differ from the true population proportion p, and this upper limit holds for at least 95% of all random samples. To express results in terms of percents instead of proportions, simply multiply everything by 100.

Conservative (approximate 95%) Margin of Error = $1/\sqrt{n}$

where *n* is the sample size.

We will see where this formula for the conservative margin of error comes from in Chapter 10 when we discuss in more detail confidence intervals for a population proportion. For now we will consider an **approximate 95% confidence interval for a population proportion** to be given by:

Approximate 95% Confidence Interval for *p*:

sample proportion $\pm\ 1/\sqrt{n}$ or expressed as $\hat{p} \pm 1/\sqrt{n}$

Try It!

Louis Harris and Associates Inc. conducted a poll of 1,250 adults to determine *How Americans Grade the School System.* The first question in the survey was: *In general, how would you rate the quality of American public schools?*

Frequency Distribution of School Quality Responses

Excellent	462
Pretty Good	288
Only Fair	225
Poor	225
Not Sure	50

a. What type of response variable is *school quality*?

b. What graph is appropriate to summarize the distribution of this variable?

c. What proportion of sampled adults rated the quality of public schools as excellent?

d. What is the conservative 95% margin of error for this survey?

e. Give an approximate 95% (conservative) confidence interval for the population proportion of all adults that rate the quality of public schools as excellent.

Interpretation Note:
Does the interval in part (e) of 34.2% to 39.8% actually contain the population proportion of all adults that rate the quality of public schools as excellent?

It either does or it doesn't, but we don't know because we don't know the value of the population proportion. (And if we did know the value of p then we would not have taken a sample of 1250 adults to try to estimate it).

The 95% confidence level tells us that in the long run, this procedure will produce intervals that contain the unknown population proportion p about 95% of the time.

f. ***Bonus #1:*** What (approximate) sample size would be necessary to have a (conservative 95%) margin of error of 2%?

Check your answer out against Table 5.1 from the text page 153.

g. ***Bonus #2:*** How does the margin of error for a sample of size 1000 from a population of 30,000 compare to the margin of error for a sample of size 1000 from a population of 100,000?

5.3 and 5.4 Sampling Methods

There are good sampling designs and poor ones.

Poor: volunteer, self-selected, convenience samples, often biased in favor of some items over others.

Good: involve random selection, giving all items a non-zero change of being selected.

Most of our inference methods require the data be considered a ...

____________________ ________________________.

This implies that the responses are to be ***independent and identically distributed (iid)***. We will make this more formal later after probability, but here are the basic ideas between these two properties.

Independent =the response you will obtain from one individual
the response you will get from another individual.

Identically distributed = all of the responses __________________________.

Many sampling designs are discussed in your text (SRS, stratified, cluster, etc). We will not cover the details of these various methods, nor work with a random number table. However, we will expect you to think about whether the data available can be considered a random sample, based on the fundamental rule for using data for inference from Chapter 5.

We will also discuss various graphs that sometimes can be used for checking assumptions, one of which is a time plot for assessing the identically distributed property of a random sample (if the data are collected over time).

5.5 Difficulties and Disasters in Sampling

This section presents some of the problems that can arise even when a sampling plan has been well designed. It talks about sampling from the wrong population, relying on volunteer response, and meaningless polls. Read over these few pages – especially Case Study 5.1 on page 166.

5.6 How to Ask Survey Questions

The wording and presentation of questions can significantly influence the results of a survey. This section presents seven pitfalls that are possible sources of response bias in a survey. Good examples!

Asking the Uninformed (page 168)

People do not like to admit that they don't know what you are talking about when you ask them a question. Crossen (1994, p. 24) gives an example: "When the American Jewish Committee studied Americans' attitudes toward various ethnic groups, almost 30% of the respondents had an opinion about the fictional Wisians, rating them in social standing above a half-dozen other real groups, including Mexicans, Vietnamese and African blacks."

Try It! Consider the following two questions:

1. *Considering that research has shown that exposure to cigarette smoke is harmful, do you think smoking should be allowed in all public restaurants or not?*
2. *Considering it is not against the law to smoke, do you agree that smoking should be allowed in all public restaurants?"*

Here are the two results:

- *30% favored banning smoking*
- *70% favored banning smoking*

Which question (1 or 2) produced the 30%, which the 70%?

A more neutral and unbiased question might be:
Do you believe smoking should or should not be allowed in all public restaurants?

Additional Notes

A place to ... jot down questions you may have and ask during office hours, take a few extra notes, write out an extra practice problem or summary completed in lecture, create your own short summary about this chapter.

Stat 250 Gunderson Lecture Notes
Chapter 6: Gathering Useful Data for Examining Relationships

A knowledge of statistics is like a knowledge of foreign languages or of algebra; it may prove of use at any time under any circumstances.

Bowley, Arthur L. Elements of Statistics *Part I, Chpt I (p. 4)* © FAIR USE

In Chapter 6 we focus on ways to collect data in order to examine relationships between variables. We will learn that the way in which a study is conducted affects our ability to support whether a cause-and-effect relationship exists.

6.1 Speaking the Language of Research Studies

Two Basic Types of Research Studies: Observational or Experimental

Definitions:

Observational Studies: The researchers simply observe or measure the participants (about opinions, behaviors, or outcomes) and do not assign any treatments or conditions. Participants are not asked to do anything differently.

Experiments: The researchers manipulate something and measure the effect of the manipulation on some outcome of interest. Often participants are ***randomly assigned*** to the various conditions or treatments.

Most studies, either observational or experimental, are interested in learning of the effect of one variable (called **explanatory variable**) on another variable (called **response** or **outcome variable**).

A **confounding variable** is a variable that both affects the response variable and also is related to the explanatory variable. The effect of a confounding variable on the response variable cannot be separated from the effect of the explanatory variable. Confounding variables might be measured and accounted for in the analysis, or they could be unmeasured **lurking variables**. Confounding variables are especially a problem in observational studies. Randomized experiments help control the influence of confounding variables.

From Utts, Jessica M. and Robert F. Heckard. Mind on Statistics, Fourth Edition. 2012. Used with permission.

Try It! Student's Health Study

A researcher at the University of Michigan believes that the number of times a student visits the Student Health Center (SHC) is strongly correlated with the student's type of diet and their amount of weekly exercise. The researcher selected a simple random sample of 100 students from a total of 3,568 students that visited SHC last month and first recorded the number of visits made to the SHC for each selected student over the previous 6 months. After recording the number of visits, he looked into their records and classified each student according to the type of diet (Home-Cooked Food / Fast Food) and the amount of exercise (None / Twice a Week / Everyday).

a. Is this an observational study or a randomized experiment?

b. What are the explanatory and response variables?

Try It! External Clues Study

A study examined how external clues influence student performance. Undergraduate students were randomly assigned to one of four different forms for their midterm exam. Form 1 was printed on blue paper and contained difficult questions, while Form 2 was also printed on blue paper but contained simple questions. Form 3 was printed on red paper, with difficult questions, and Form 4 was printed on red paper with simple questions. The researchers were interested in the impact that color and type of question had on exam score (out of 100 points).

a. This research is based on: ***an observational study | a randomized experiment***

b. Complete the following statements by circling the appropriate answer.

   i. The color of the paper is a(n) ***response*** ***explanatory*** variable

      and its type is (circle one) ***categorical*** ***quantitative***.

   ii. The exam score is a(n) ***response*** ***explanatory*** variable

      and its type is (circle one) ***categorical*** ***quantitative***.

c. Fill in the blank. Suppose the students in the "blue paper" group were mostly upper-classmen and the students in the "red paper" group
   were mostly first and second-year students. The variable "class rank" is an example of a(n) ______________________________ variable.

Sections 6.2 – 6.3 discuss key ideas for designing a good experiment and a good observational study. Section 6.4 presents a list of difficulties and disasters in experiments and observational studies and provides good examples of each.

You need to know about these ideas (control groups, placebos, blinding, matched-pair designs, interacting variables, biases like the Hawthorne effect, and more – so read these sections!

Where are we going?

- We have a population (a basket) that we cannot examine but we want to learn something about it - so we will take a sample - preferably it will be a random sample.
- We will use the sample to *estimate* the things we wanted to know about the population - we will use the sample results to test theories about the population and make some decisions.
- Since the sample is just a part of the population there will be some uncertainty about the estimates and decisions we make. To measure and quantify that uncertainty we turn to **PROBABILITY**! So we turn next to Chapter 7. Chapters 3 and 4 will be combined with later chapters that also cover those ideas but in more detail.

Additional Notes

A place to … jot down questions you may have and ask during office hours, take a few extra notes, write out an extra practice problem or summary completed in lecture, create your own short summary about this chapter.

Stat 250 Gunderson Lecture Notes
Chapter 7: Probability

Chance favors prepared minds. -- *Louis Pasteur* © FAIR USE

Many decisions that we make involve uncertainty and the evaluation of probabilities.

7.1 Random Circumstance
7.2 Interpretations of Probability

Example: Roll a fair die → possible outcomes = { }

Before you roll the die do you know which one will occur?

What is the probability that the outcome will be a '4'? _________ Why?

A few ways to think about PROBABILITY:

(1) Personal or Subjective Probability

P(A) = the degree to which a given individual believes that the event A will happen.

(2) Long term relative frequency

P(A) = proportion of times 'A' occurs if the random experiment (circumstance) is repeated many, many times.

(3) Basket Model

P(A) = proportion of balls in the basket that have an 'A' on them.

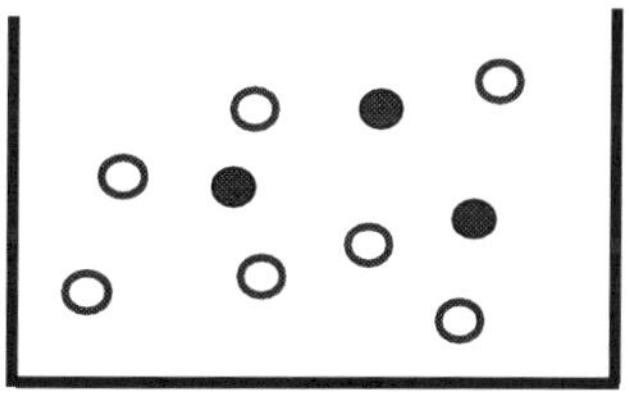

10 balls in the basket: 3 blue and 7 white
One ball will be selected at random.
What is P(blue)? ____________

Note: A probability statement ***IS NOT*** a statement about __________________.

It ***IS*** a statement about ___________________________.

7.3 Probability Definitions and Relationships
7.4 Basic Rules for Finding Probability

There is a lot you can learn about probability. One basic rule to always keep in mind is that the probability of any outcome is always between 0 and 1. Now, there are entire courses devoted just to studying probability. But this is a Statistics class. So rather than start with a list of definitions and formulas for finding probabilities, let's just do it through an example so you can see what ideas about probability we need to know for doing statistics.

Example: Shopping Online

Many Internet users shop online. Consider a population of 1000 customers that shopped online at a particular website during the past holiday season and their results regarding whether or not they were satisfied with the experience and whether or not they received the products on time. These results are summarized below in table form. Using the idea of probability as a proportion, try answering the following questions.

	On Time	**Not On Time**	***Total***
Satisfied	800	20	820
Not Satisfied	80	100	180
Total	880	120	1000

a. What is the probability that a randomly selected customer was satisfied with the experience?

b. What is the probability that a randomly selected customer was *not* satisfied with the experience?

c. What is the probability that a randomly selected customer was both satisfied *and* received the product on time?

d. What is the probability that a randomly selected customer was either satisfied *or* received the product on time?

e. *Given* that a customer did receive the product on time, what is the probability that the customer was satisfied with the experience?

f. *Given* that a customer did *not* receive the product on time, what is the probability that the customer was satisfied with the experience?

***Note**: We stated the above 1000 customers represented a population. If results were based on a sample that is representative of a larger population, then the observed sample proportions would be used as approximate probabilities for a randomly selected person from the larger population.*

Great job! You just computed probabilities using many of the basic probability rules or formulas summarized below and also found in your textbook.

- **Complement rule** $P(A^C) = 1 - P(A)$
- **Addition rule** $P(A \text{ or } B) = P(A) + P(B) - P(A \text{ and } B)$
- **Multiplication rule** $P(A \text{ and } B) = P(A)P(B \mid A)$
- **Conditional Probability** $P(A \mid B) = \dfrac{P(A \text{ and } B)}{P(B)}$

You did not need the formulas themselves but instead used intuition and approaching it as some type of proportion. Let's see how your intuition and the above formulas really do connect.

In part b you found the probability of "<u>NOT</u> being satisfied", which is the complement of the event "being satisfied", so the answer to part b is the complement of the probability you found in part a.

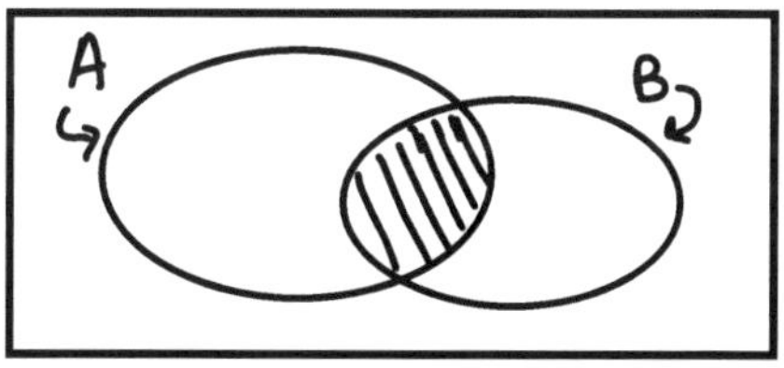

In part c, there was a key word of "<u>AND</u>" in the question being asked. The "AND" is just the intersection, or the overlapping part; the outcomes that are in common. The picture at the right show an intersection between the event A and B. In a table, the counts that are in the middle are the "AND" counts; there were 800 (out of the 1000 customers) that were *both* satisfied AND received the product on time. There is a *multiplication* formula above for finding probabilities of the AND or intersection of two events, but <u>we did not even need to apply it</u>; as a table presentation of counts provides "AND" counts directly.

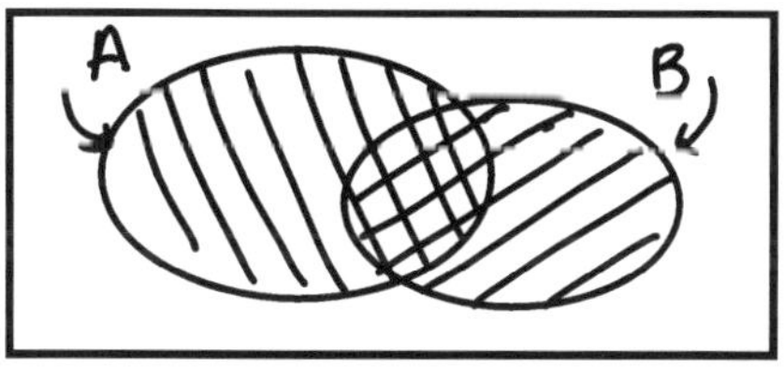

In part d, there was a key word of "OR" in the question being asked. The "OR" is union, the outcomes that are in either one or the other (including those that are in both). The picture at the right show an union between the event A and B. Notice that if you start with all of the outcomes that are in A and then add all of the outcomes that are in B, you have double counted the outcomes that are in the overlap. So the *addition* formula above shows you need to subtract off the intersecting probability once to correct for the double counting. From the table, you could either add up the separate counts of 800 + 20 + 80; or start with the 820 that were satisfied and add the 880 that received it on time and then subtract the 800 that were in both sets; to get the 900 in all that were either satisfied *or* received the product on time.

Finally, parts e and f were both conditional probabilities. In part e you were first told to consider only the 880 customers that received the product on time, and out of these find the probability (or proportion) that were satisfied. There were 800 out of the 880 that were satisfied. The picture below shows the idea of a conditional probability formula above for P(A | B), read as the probability of A given B has occurred. If we know B has occurred, then only look at those items in the event B. The event B, shaded at the right, is our new 'base' (and thus is in the denominator of the formula). Now out of those items in B, we want to find the probability of A. The only items in A that are on the set B are those in the overlap or intersection. So the *conditional* formula above shows you count up those in the "A and B" and divide it by the base of "B".

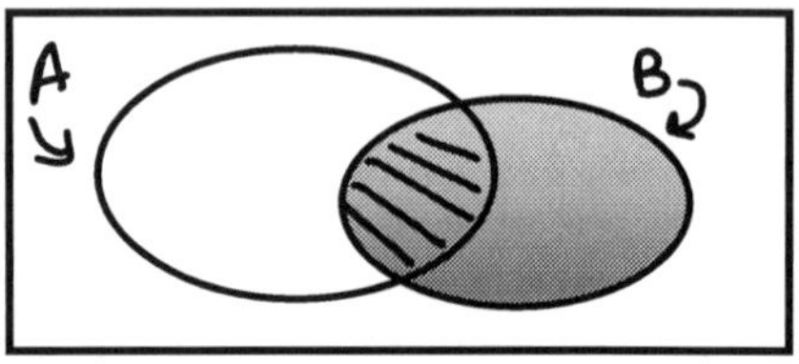

Try It! Go back to parts a to f and add the corresponding shorthand probability notation of what you actually found; e.g. **P(satisfied), P(satisfied | on time)** next to each answer.

Now there are a couple of useful situations that can make computing probabilities easier.

> ***Definition:***
> Two events A, B are ***Mutually Exclusive*** (or ***Disjoint***) if ...
> **they do not contain any of the same outcomes.**
> **So their intersection is empty.**

From Utts, Jessica M. and Robert F. Heckard. Mind on Statistics, Fourth Edition. 2012. Used with permission.

We can easily *picture* disjoint events because the definition is a property about the sets themselves.

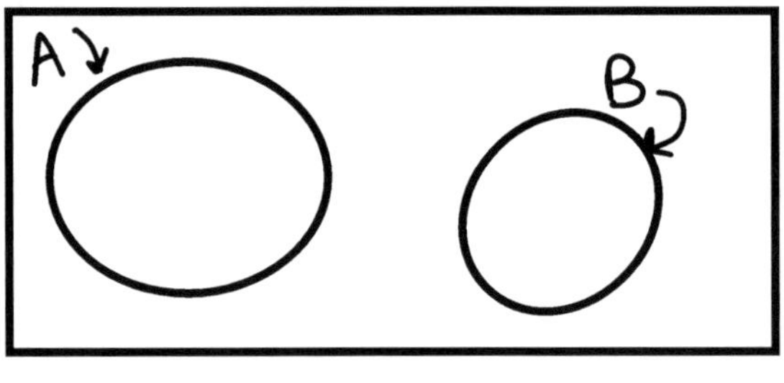

If A, B are disjoint, then P(A and B) = 0. If there are no items in the overlapping part, then man of the probability results will simplify. For example, the additional rule for disjoint events P(A or B) = P(A) + P(B).

Another important situation in statistics occurs when the two events turn out to be ***independent***.

> ***Definition:***
> Two events A, B are said to be **independent** if knowing that one will occur (or has occurred) does not change the probability that the other occurs. In notation this can be expressed as **P(A|B) = P(A).**

From Utts, Jessica M. and Robert F. Heckard. Mind on Statistics, Fourth Edition. 2012. Used with permission.

This expression P(A|B) = P(A) tells us that knowing the event B occurred does not change the probability of the event A happening. Now it works the other way around too, namely, if A and B are independent events, then P(B|A) = P(B).

As a result of this independence definition, we could show that the multiplication rule for independent events reduces to P(A and B) = P(A)P(B).

Finally, this rule can also be extended. If three events A, B, C are all independent then P(A and B and C) = P(A)P(B)P(C).

So let's apply these two new concepts to our online shopping example.

Back to the Shopping Online Example

Below are results for a population of 1000 customers that shopped online at a particular website during the past holiday season. Recorded was whether or not they were satisfied with the experience and whether or not they received the products on time.

	On Time	**Not On Time**	***Total***
Satisfied	800	20	820
Not Satisfied	80	100	180
Total	880	120	1000

g. Are being satisfied with the experience and receiving the product on time *mutually exclusive* (*disjoint*)? Provide support for your answer.

h. Are being satisfied with the experience and receiving the product on time statistically *independent*? Provide support for your answer.
Hint: go back and compare your answers to parts a, e, and f.

Try It! Elderly People

Suppose that in a certain country, 10% of the elderly people have diabetes. It is also known that 30% of the elderly people are living below poverty level and 5% of the elderly population falls into both of these categories. *At the right is a diagram for these events. Do the probabilities make sense to you?*

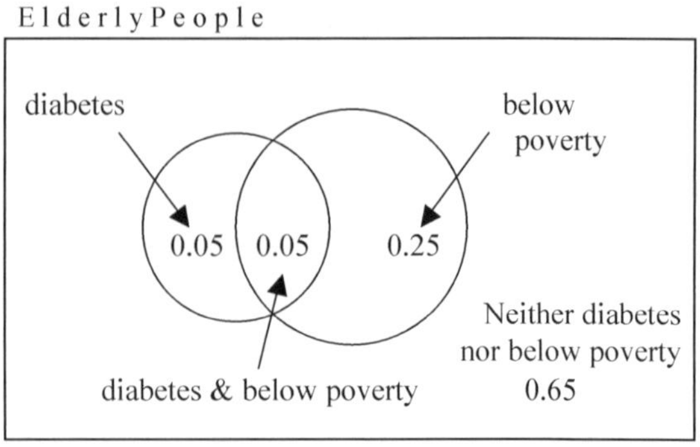

a. What is the probability that a randomly selected elderly person is not diabetic?

b. What is the probability that a randomly selected elderly person is either diabetic *or* living below poverty level?

c. Given a randomly selected elderly person is living below poverty level, what is the probability that he or she has diabetes?

d. Since knowing an elderly person lives below the poverty level (circle one)

DOES **DOES NOT** change the probability that they are diabetic, the two events of living below the poverty level and being diabetic (circle one)

ARE **ARE NOT** independent.

In this next example, you are not asked to determine if two events are independent, but rather put independence to use.

Try It! Blood Type

About 1/3 of all adults in the United States have type O+ blood.
Suppose three adults will be randomly selected.

Hint: randomly selected implies the results should be_______________________.

What is the probability that the **first** selected adult will have type O+ blood?

What is the probability that the **second** selected adult will have type O+ blood?

What is the probability that **all three** will have type O+ blood?

What is the probability that **none** of the three will have type O+ blood?

What is the probability that **at least one** will have type O+ blood?

Some final notes…

I. Sampling with and without Replacement

> ***Definitions:***
> A sample is drawn **with replacement** if individuals are returned to the eligible pool for each selection.
> A sample is drawn **without replacement** if sampled individuals are not eligible for subsequent selection.

From Utts, Jessica M. and Robert F. Heckard. Mind on Statistics, Fourth Edition. 2012. Used with permission.

If sampling is done with replacement, the Extension of Rule 3b holds. If sampling is done without replacement, probability calculations can be more complicated because the probabilities of possible outcomes at any specific time in the sequence are conditional on previous outcomes.

If a sample is drawn from a very large population, the distinction between sampling with and without replacement becomes unimportant. In most polls, individuals are drawn without replacement, but the analysis of the results is done as if they were drawn with replacement. The consequences of making this simplifying assumption are negligible.

II. Sometimes students confuse the mutually exclusive with independence.

Check the definitions.

- The definition for two events to be ***disjoint*** (mutually exclusive) was based on a **<u>SET</u>** property.
- The definition for two events to be ***independent*** is based on a **<u>PROBABILITY</u>** property.

You need to check if these definitions hold when asked to assess if two events are disjoint, or if two events are independent.

Mutually Exclusive  **Independence**

If two events are mutually exclusive then we know that P(A and B) = 0.

This also implies that P(A|B) is equal to 0 (if the two events are disjoint and B did occur, then the chance of A occurring is 0).

So P(A|B) (which is 0) will not be equal to P(A) if the events are disjoint.

III. Probability rules summary

Below is a summary of the key probability results you need to understand and be able to use.

- **Complement rule** $P(A^C) = 1 - P(A)$
- **Mutually Exclusive (disjoint) Events:**
  The events A, B are disjoint if "A and B" is the empty set.
  Thus, P(A and B) = 0.
- **Addition Rule (general)** $P(A \text{ or } B) = P(A) + P(B) - P(A \text{ and } B)$
  If A, B are disjoint, we have $P(A \text{ or } B) = P(A) + P(B)$
- **Conditional Probability (general)** $P(A \mid B) = \dfrac{P(A \text{ and } B)}{P(B)}$
- **Independent Events:**
  The events A, B are independent if $P(A \mid B) = P(A)$
  Equivalently, the events A, B are independent
  if $P(A \text{ and } B) = P(A)P(B)$

The Stats 250 formula card provides a more extensive list, but remember, you may not need them as you discovered in your first probability example with the online customers.

Additional Notes

A place to ... jot down questions you may have and ask during office hours, take a few extra notes, write out an extra practice problem or summary completed in lecture, create your own short summary about this chapter.

Stat 250 Gunderson Lecture Notes
Chapter 8: Random Variables

All models are wrong; some models are useful. -- *George Box*

Patterns make life easier to understand and decisions easier to make. In Chapter 2 we discussed the different types of data or variables and how to turn the data into useful information with graphs and numerical summaries. Having some notion of probability from the previous chapter, we can now view the variables as "random variables" – the numerical outcomes of a random circumstance. We will look at the pattern of the distribution of the values of a random variable and we will see how to use the pattern to find probabilities. These patterns will serve as models in our inference methods to come.

8.1 What is a Random Variable?

Recall in our discussion on probability we started out with some random circumstance or experiment that gave rise to our set of all possible outcomes *S*. We developed some rules for calculating probabilities about various events. Often the events can be expressed in terms of a "random variable" taking on certain outcomes. Loosely, this random variable will represent the value of the variable or characteristic of interest, but *before we look*. Before we look, the value of the variable is not known and could be any of the possible values with various probabilities, hence the name of a "random" variable.

Definition:
A ***random variable*** assigns a number to each outcome of a random circumstance, or, equivalently, a random variable assigns a number to each unit in a population.

From Utts, Jessica M. and Robert F. Heckard. Mind on Statistics, Fourth Edition. 2012. Used with permission.

We will consider ***two broad classes*** of random variables: **discrete** random variables and **continuous** random variables.

Definitions:
A ***discrete random variable*** can take one of a countable list of distinct values.
A ***continuous random variable*** can take any value in an interval or collection of intervals.

From Utts, Jessica M. and Robert F. Heckard. Mind on Statistics, Fourth Edition. 2012. Used with permission.

Try It! Discrete or Continuous

A car is selected at random from a used car dealership lot. For each of the following characteristics of the car, decide whether the characteristic is a continuous or a discrete random variable.

a. Weight of the car (in pounds).

b. Number of seats (maximum passenger capacity).

c. Overall condition of car (1 = good, 2 = very good, 3 = excellent).

d. Length of car (in feet).

In statistics, we are interested in the **distribution of a random variable** and we will use the distribution to compute various probabilities. The probabilities we compute (for example, *p*-values in testing theories) will help us make reasonable decisions.

So just what is the distribution of a random variable? Loosely, it is a model that shows us what values are possible for that particular random variable and how often those values are expected to occur (i.e. their probabilities). The model can be expressed as a function, table, picture, depending on the type of variable it is.

In sections 8.2, 8.3, and 8.4 we will discuss discrete random variables and their models. We first work with the broad class of general discrete random variables and in 8.4 we will focus on a **particular family of discrete random variables** called the **Binomial**. The Binomial random variable arises in situations where you are counting the number of *successes* that occur in a sample. Sections 8.5 and 8.6 look at properties for continuous random variables and then spend more time studying the **family of uniform random variables and normal random variables.** In later chapters you will be introduced to more models for continuous random variables that are primarily used in statistical testing problems. Below is a summary of the types of random variables we will work with in this course.

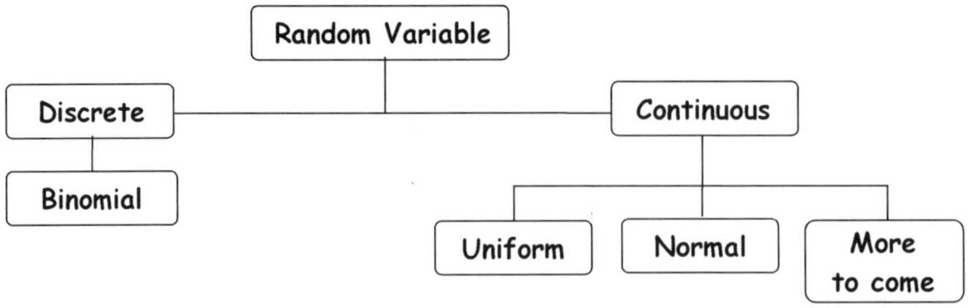

Technical Note: Sometimes a random variable fits the technical definition of a discrete random variable but it is more convenient to treat it, that is, model it, as if it were continuous. In section 8.7 we will see when it is reasonable to model a discrete binomial random variable as being approximately normal. Finally Section 8.8 will show us how to model sums and differences of random variables.

Some general notes about random variables are:

- random variables will be denoted by capital letters (*X, Y, Z*);
- outcomes of random variables are represented with small letters (*x, y, z*).

So when we express probabilities about the possible value of a random variable we use the capital letter. For example, the probability that a random variable takes on the value of 2 would be expressed as $P(X = 2)$.

8.2 General Discrete Random Variables

A **discrete** random variable, X, is a random variable with a finite or countable number of possible outcomes. The probability notation your text uses for a Discrete Random Variable is given next:

> ***Discrete Random Variable:***
> X = the random variable.
> k = a number that the discrete random variable could assume.
> $P(X = k)$ is the probability that the random variable X equals k.

The **probability distribution function (pdf) for a discrete random variable *X*** is a table or rule that assigns probabilities to the possible values of the *X*.
One way to show the distribution is through a table that lists the possible values and their corresponding probabilities:

Value of *X*	x_1	x_2	x_3	...
Probability	p_1	p_2	p_3	...

- **Two conditions** that must always apply to the probabilities for a discrete random variable are:
  Condition 1: The sum of all of the individual probabilities must equal 1.
  Condition 2: The individual probabilities must be between 0 and 1.

- A ***probability histogram*** or better yet, a ***probability stick graph***, can be used to display the distribution for a discrete random variable.
  The *x*-axis represents the values or outcomes.
  The *y*-axis represents the probabilities of the values or outcomes.

- The **cumulative distribution function (cdf) for a discrete random variable *X*** is a table or rule that provides the probabilities $P(X \leq k)$ for any real number *k*. Generally, the term cumulative probability refers to the probability that *X* is ***less than or equal to*** a particular value.

Try It! Psychology Experiment

A psychology experiment on the behavior of young children involves placing a child in a designated area with five different toys. Over a fixed time period various observations are made. One response measured is the number of toys the child plays with.

Psychology Experiment *continued*

Based on many results, the probability distribution on the next page was determined for the general discrete random variable X = number of toys played with by children (during a fixed time period).

X = # toys	0	1	2	3	4	5
Probability	0.03	0.16	0.30	0.23	0.17	

a. What is the missing probability $P(X = 5)$?

b. Graph this discrete probability distribution function for X.

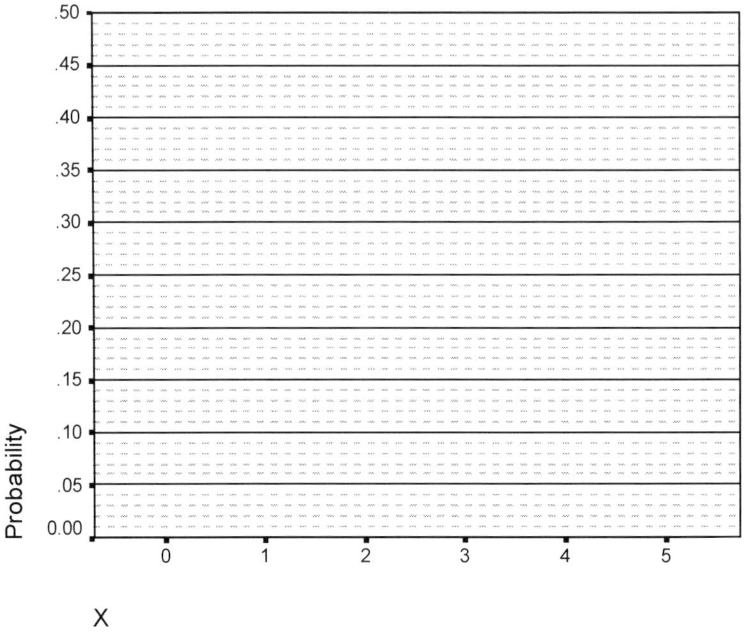

c. What is the probability that a child will play with ***at least*** 3 toys?

d. Given the child has played with at least 3 toys,
what is the probability that he/she will play with all 5 toys?

e. Finish the table below to provide the cumulative distribution function of X.

X = # toys	0	1	2	3	4	5
Cum Probability ***$P(X \leq k)$***	0.03	0.03+0.16 = 0.19	0.03+0.16+0.30 = 0.49			

8.3 Expectations for Random Variables

Just as we moved from summarizing a set of data with a graph to numerical summaries, we next consider computing the mean and the standard deviation of a random variable. The mean can be viewed as the expected value over the long run (in many repetitions of the random circumstance) and the standard deviation can be viewed is approximately the average distance of the possible values of X from its mean.

> ***Definition:***
> The ***expected value*** of a random variable is the mean value of the variable X in the sample space, or population, of possible outcomes. *Expected value*, denoted by $E(X)$, can also be interpreted as the mean value that would be obtained from an infinite number of observations on the random variable.

From Utts, Jessica M. and Robert F. Heckard. Mind on Statistics, Fourth Edition. 2012. Used with permission.

Motivation for the expected value formula ...

Consider a population consisting of 100 families in a community. Suppose that 30 families have just 1 child, 50 families have 2 children, and 20 families have 3 children. What is the mean or average number of children per family for this population?

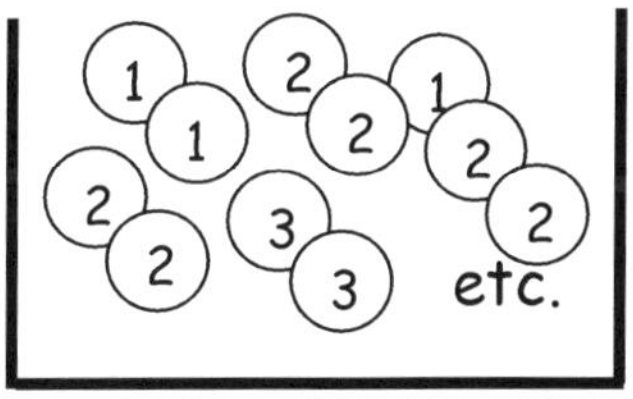

Population of 100 families

Mean = (sum of all values)/100
= [1(30) + 2(50) + 3(20)]/100
= **1**(30/100) + **2**(50/100) + **3**(20/100)
= 1(0.30) + 2(0.50) + 3(0.20)
= 1.9 children per family

Mean = Sum of (value x **probability of that value)**

> ***Definitions:***
> **Mean and standard deviation of a discrete random variable**
> Suppose that X is a discrete random variable with possible values $x_1, x_2, x_3, \ldots$ occurring with probabilities $p_1, p_2, p_3, \ldots$, then
> the expected value (or mean) of X is given by $\mu = E(X) = \sum x_i p_i$
> the *variance* of X is given by $V(X) = \sigma^2 = \sum (x_i - \mu)^2 p_i$
> and so the *standard deviation* of X is given by $\sigma = \sqrt{\sum (x_i - \mu)^2 p_i}$
>
> The sums are taken over all possible values of the random variable X.

From Utts, Jessica M. and Robert F. Heckard. Mind on Statistics, Fourth Edition. 2012. Used with permission.

Try It! Psychology Experiment

Recall the probability distribution for the discrete random variable X = number of toys played with by children.

X = # toys	0	1	2	3	4	5
Probability	0.03	0.16	0.30	0.23	0.17	0.11

a. What is the expected number of toys played with?

Note: The expected value may not be a value that is ever expected on a single random outcome. Instead, it is the average over the long run.

b. What is the standard deviation for the number of toys played with?

c. Complete the interpretation of this standard deviation (in terms of an average distance):

On average, the number of toys played with vary by about ____________

from the mean number of toys played with of ___________.

8.4 Binomial Random Variables

An important class of discrete random variables is called the **Binomial Random Variables**.

A binomial random variable is that it ***COUNTS*** the number of times a certain event occurs out of a particular number of observations or trials of a random experiment.

Examples of Binomial Random Variables:

- The number of girls in six independent births.
- The number of tall men (over 6 feet) in a random sample of 30 men from a large male population.

A **binomial experiment** is defined by the following conditions:

1. There are *n* "trials", *n* is determined in advance and is not a random value.
2. There are two possible outcomes on each trial,
   called "success" (S) and "failure" (F).
3. The outcomes are independent from one trial to the next.
4. The probability of a "success" remains the same from one trial to the next, and this probability is denoted by *p*. The probability of a "failure" is $1 - p$ for every trial.

A **binomial random variable** is defined as

***X* = number of successes in the *n* trials of a binomial experiment.**

Table 8.4 (page 276) has good examples and provides some good points to keep in mind when trying to decide whether a random variable fits the binomial description (read them).

Try It! Are the Conditions Right for Binomial?

a. Observe the sex of the next 50 children born at a local hospital.
   X = number of girls

b. A ten-question quiz has five True-False questions and five multiple-choice questions, each with four possible choices. A student randomly picks an answer for every question.
   X = number of answers that are correct.

c. Four students are randomly picked without replacement from large student body listing of 1000 women and 1000 men.
   X = number of women among the four selected students.

What if the student body listing consisted only of 10 women and 10 men?

***Rule of Thumb*: population at least 10 times as large as the sample → ok!**

The Binomial Formula

We will develop the formula together using our probability knowledge. Suppose that of the online shoppers for a particular website that start filling a shopping cart with items, 25% actually make a purchase (complete a transaction). We have a random sample of 10 such online shoppers.

If the stated rate is true, what is the probability that ...

... all 10 shoppers will actually make a purchase?

... none of the shoppers will make a purchase?

... just 1 shopper will make a purchase?

Do you know that with only the probability knowledge from chapter 7, you just calculated three binomial probabilities that are based on the following formula?

The *binomial distribution:*

Probability of exactly *k* successes in *n* trials ...

$$P(X = k) = \binom{n}{k} p^k (1-p)^{n-k} \qquad \text{for } k = 0,1,2,\ldots,n$$

where $\binom{n}{k} = \dfrac{n!}{k!(n-k)!}$ *(this represents the # of ways to select k items from n)*

Try it! The first part ...

You can think of the computation of $\binom{n}{k}$ in the following way ...

Suppose you had n friends, how many ways could you invite k to dinner? The ones "at the ends" are easy to do without even using the formula or a calculator. Your calculator is likely to have this complete function or at least a factorial ! option. On many calculators this combinations function is found under the math → probability menu and expressed as nCr.

1. $\binom{10}{0} =$

2. $\binom{10}{10} =$

3. $\binom{10}{1} =$

4. $\binom{10}{9} =$

5. $\binom{10}{2} =$

Try it! Finding Binomial Probabilities

Recall we have a random sample of $n = 10$ online shoppers from a large population of such shoppers and that $p = 0.25$ is the population proportion who actually make a purchase.

a. What is the probability of selecting ***exactly one*** shopper who actually makes a purchase?

b. What is the probability of selecting ***exactly two*** shoppers who actually make a purchase?

c. What is the probability of selecting ***at least one*** shopper who actually makes a purchase?

d. How many shoppers in your random sample of size 10 would you ***expect*** to actually make a purchase?

In the previous question (part d), you just computed the **mean of a binomial distribution**.

> If X has the **binomial distribution Bin(n, p)** then
>
> **Mean** of X is $\mu = E(X) = \boldsymbol{np}$
>
> **Standard Deviation** of X, is $\sigma = \sqrt{np(1-p)}$

Try it! More Work with the Binomial

Suppose that about 10% of Americans are left-handed. Let X represent the number of left-handed Americans in a random sample of 12 Americans.

Then X has a ______________________________ distribution (*be as specific as you can*).

Note that the mean or expected number of left-handed Americans in such a random sample would be $\mu = np = 12(0.10) = 1.2$. The standard deviation (reflecting the variability in the results from the mean across many such random samples) is $\sigma = \sqrt{np(1-p)} = \sqrt{12(0.10)(0.90)} = 1.04$.

a. What is the probability that the sample contains 2 or fewer left-handed Americans?

b. Suppose a random sample of 120 Americans had been taken instead of just 12. So now X has a Bin(n = 120, p = 0.10) model. The mean or expected number of left-handed Americans in a random sample is $\mu = np = 120(0.10) = 12$. The standard deviation $\sigma = \sqrt{np(1-p)} = \sqrt{120(0.10)(0.90)} = 3.29$.

   So how might you try to find the probability that a random sample of 120 Americans would result in 20 or fewer left-handed Americans? Note that 2 out of n = 12 is 16.67% and that 20 out of n = 120 is also equal to 16.67%.

8.5 General Continuous Random Variables

A **continuous random variable**, X, takes on all possible values in an interval (or a collection of intervals). The way that we determine probabilities for continuous random variables differs in one important respect from how we determine probabilities for discrete random variables. For a discrete random variable, we can find the probability that the variable X exactly equals a specified value. We can't do this for a continuous random variable. Instead, we are only able to find the probability that X could take on values in an interval. We do this by determining the corresponding area under a ***curve*** called the probability density function of the random variable.

In Chapter 2, we summarized the general shapes of distributions of a quantitative response that often arise with real data. The shape of a distribution was found by drawing a smooth ***curve*** that traces out the overall pattern that is displayed in a histogram. With a histogram, the area of each rectangle is proportional to the frequency or count for each class. The curve also provides a visual image of proportion through its area. If we could get the equation of this smoothed curve, we would have a simple and somewhat accurate summary of the distribution of the response.

The picture at the right shows a smoothed curve that is symmetric and bell shaped, even though the underlying histogram is only approximately symmetric. If the data came from a representative sample, the smooth curve could serve as a model, that is, as the probability distribution for the continuous response for the population.

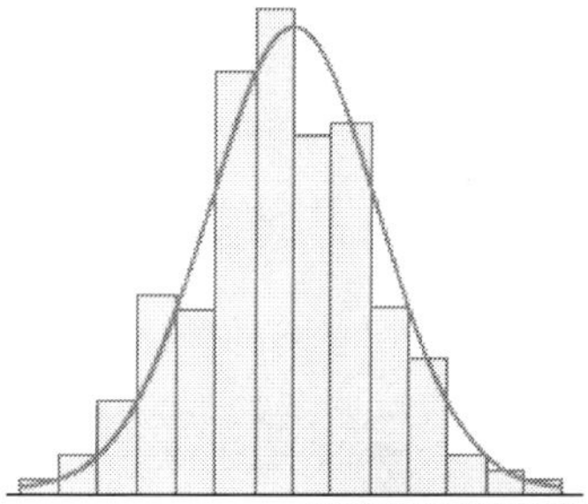

So the **probability distribution of a continuous random variable** is described by a **density curve**. The probability of an event is the area under the curve for the values of X that make up the event. The probability model for a continuous random variable assigns probabilities to intervals.

> ***Definition:***
> A curve (or function) is called a ***Probability Density Curve*** if:
> 1. It lies on or above the horizontal axis.
> 2. Total area under the curve is equal to 1.

From Utts, Jessica M. and Robert F. Heckard. Mind on Statistics, Fourth Edition. 2012. Used with permission.

KEY IDEA: AREA under a density curve over a range of values corresponds to the PROBABILITY that the random variable *X* takes on a value in that range.

Try It! Some Probability Density Curves

I. A density curve for modeling income for employed adults (in $1000s) for a city.

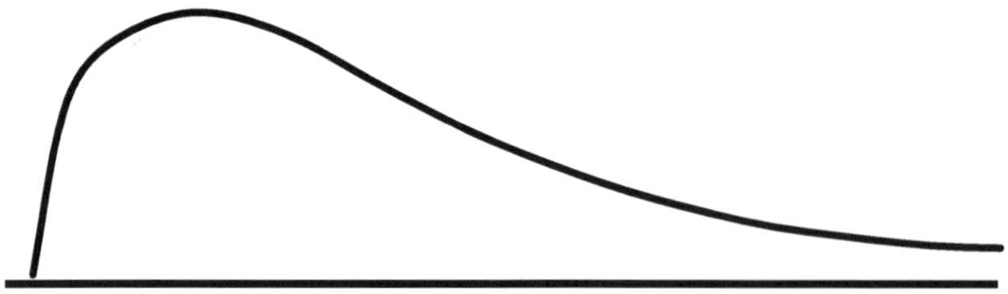

How would you use the above density curve to estimate the probability of a randomly selected employed adult from this city having an income between $30,000 and $40,000?

II. Consider the following curve:

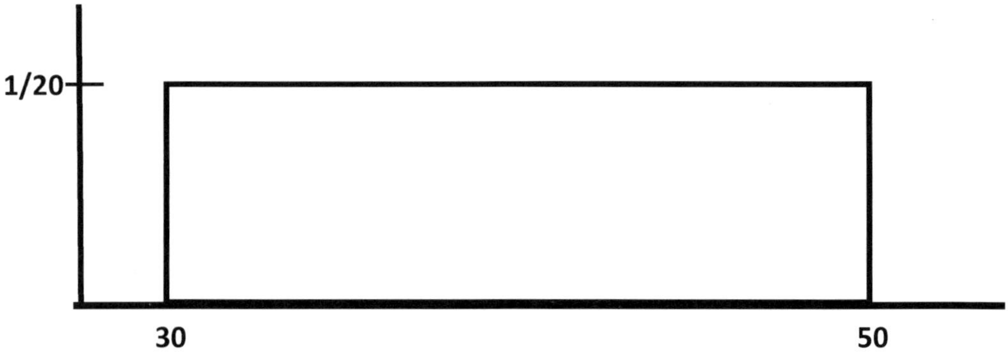

a. Is this a density curve? Why?

b. If yes, find the probability of observing a response that is less than 35.

c. What does the value of 35 correspond to for this distribution?

Try it! Checkout time at a store

Let X be the checkout time at a store, which is a random variable that is uniformly distributed on values between 5 and 20 minutes. That is, X is $U(5, 20)$.

a. What does the density look like? Sketch it and include a value on the y-axis.

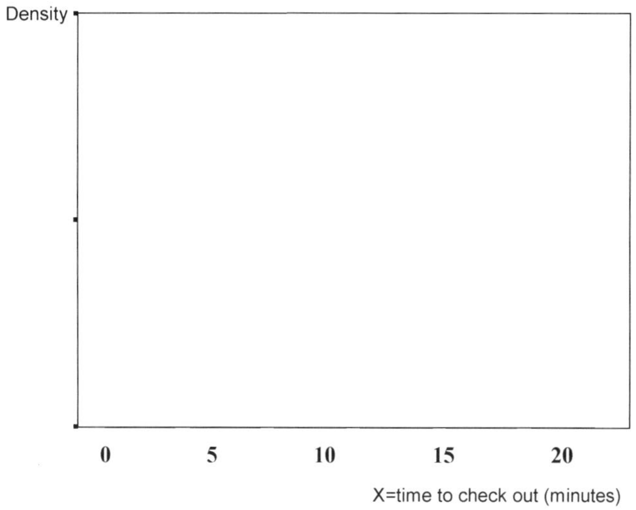

b. What is the probability a person will take more than 10 minutes to check out?

c. Given a person has already spent 10 minutes checking out, what is the probability they will take no more than 5 additional minutes to check out?

d. What is the expected time to check out at this store?

***Definition:* Mean of a continuous random variable.**

Expected Value or Mean = Balancing point of the density curve **E(X) =** μ

(Sometimes one would need calculus/integration to find it -- integral instead of sums)

From Utts, Jessica M. and Robert F. Heckard. Mind on Statistics, Fourth Edition. 2012. Used with permission.

There are many density curves that can be used as models. In our next Section 8.6 we focus on an important family of densities called the **NORMAL DISTRIBUTIONS.**

8.6 Normal Random Variables

We had our first introduction to normal random variables back in Chapter 2 as a special case of bell-shaped distributions. The family of normal distributions is very important because many variables have this shape and form approximately and many statistics that we use in our inference methods are based on sums or averages which generally have (approximately) a normal distribution.

A Normal Curve: Symmetric, bell-shaped, centered at the mean μ and its spread is determined by the standard deviation σ. In fact, the points of inflection on each side of the mean mark the values which are one standard deviation away from the mean.

Notation: If a population of measurements follows a normal curve, and if *X* is the measurement for a randomly selected individual from the population, then *X* is said to be a **normal random variable**. *X* is also said to have a **normal distribution**. Any normal random variable can be completely characterized by its mean and its standard deviation.

The variable *X* is normally distributed with mean μ and standard deviation σ is denoted by:

A $N(50,10)$ curve is sketched below. Add a $N(80,5)$ curve to this picture. Keep in mind the features of the empirical rule (68-95-99.7) which applies to a normal curve.

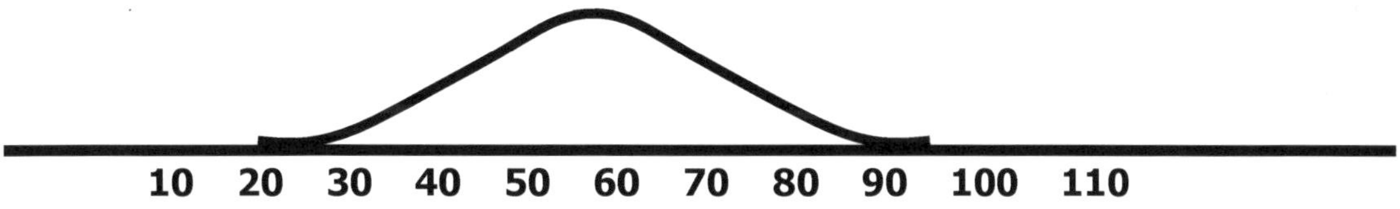

Standardized Scores:

A normal distribution is indexed by its population mean μ, and its population standard deviation σ, and denoted by $N(\mu,\sigma)$. Recall that the standard deviation is a useful "**yardstick**" for measuring how far an individual value falls from the mean. The **standardized score** or ***z*-score** is the distance between the observed value and the mean, measured in terms of number of standard deviations. Values that are above the mean have positive *z*-scores, and values that are below the mean have negative *z*-scores.

Standardized score or *z*-score: $z = \dfrac{\text{value} - \text{mean}}{\text{Standard deviation}} = \dfrac{x-\mu}{\sigma}$

Finding Probabilities for z-Scores:

Standard scores play a role in how we will find areas (and thus probabilities) under a normal curve. We simply have to convert the endpoints of the interval of interest to the corresponding standardized scores and then use a table (or computer/calculator) to find probabilities associated with these standardized scores. When we convert to standardized scores, the random variable *X* is converted to what is called the **Standard Normal Random Variable**, denoted by *Z*, and it has the $N(0,1)$ distribution.

Table A.1 provides the areas to the left for various values of Z

- The body of the table contains $P(Z \leq z^*)$, which is the probability that the value of a standard normal variable Z takes a value that is less than the specific value z^*.
- The left-most column of the table shows the algebraic sign, the digit before the decimal place, and the first decimal place for z^*.
- The second decimal place of z^* is in the column heading across the top of the table.

From Utts, Jessica M. and Robert F. Heckard. Mind on Statistics, Fourth Edition. 2012. Used with permission

Table entry for z is the area to the left of z

z	*.00*	*.01*	*.02*	*.03*	*.04*	*.05*	*.06*	*.07*	*.08*	*.09*
−3.4	.0003	.0003	.0003	.0003	.0003	.0003	.0003	.0003	.0003	.0002
−3.3	.0005	.0005	.0005	.0004	.0004	.0004	.0004	.0004	.0004	.0003
−3.2	.0007	.0007	.0006	.0006	.0006	.0006	.0006	.0005	.0005	.0005
−3.1	.0010	.0009	.0009	.0009	.0008	.0008	.0008	.0008	.0007	.0007
−3.0	.0013	.0013	.0013	.0012	.0012	.0011	.0011	.0011	.0010	.0010
−2.9	.0019	.0018	.0018	.0017	.0016	.0016	.0015	.0015	.0014	.0014
−2.8	.0026	.0025	.0024	.0023	.0023	.0022	.0021	.0021	.0020	.0019
−2.7	.0035	.0034	.0033	.0032	.0031	.0030	.0029	.0028	.0027	.0026
−2.6	.0047	.0045	.0044	.0043	.0041	.0040	.0039	.0038	.0037	.0036
−2.5	.0062	.0060	.0059	.0057	.0055	.0054	.0052	.0051	.0049	.0048
−2.4	.0082	.0080	.0078	.0075	.0073	.0071	.0069	.0068	.0066	.0064
−2.3	.0107	.0104	.0102	.0099	.0096	.0094	.0091	.0089	.0087	.0084
−2.2	.0139	.0136	.0132	.0129	.0125	.0122	.0119	.0116	.0113	.0110
−2.1	.0179	.0174	.0170	.0166	.0162	.0158	.0154	.0150	.0146	.0143
−2.0	.0228	.0222	.0217	.0212	.0207	.0202	.0197	.0192	.0188	.0183
−1.9	.0287	.0281	.0274	.0268	.0262	.0256	.0250	.0244	.0239	.0233
−1.8	.0359	.0351	.0344	.0336	.0329	.0322	.0314	.0307	.0301	.0294
−1.7	.0446	.0436	.0427	.0418	.0409	.0401	.0392	.0384	.0375	.0367
−1.6	.0548	.0537	.0526	.0516	.0505	.0495	.0485	.0475	.0465	.0455
−1.5	.0668	.0655	.0643	.0630	.0618	.0606	.0594	.0582	.0571	.0559
−1.4	.0808	.0793	.0778	.0764	.0749	.0735	.0721	.0708	.0694	.0681
−1.3	.0968	.0951	.0934	.0918	.0901	.0885	.0869	.0853	.0838	.0823
−1.2	.1151	.1131	.1112	.1093	.1075	.1056	.1038	.1020	.1003	.0985
−1.1	.1357	.1335	.1314	.1292	.1271	.1251	.1230	.1210	.1190	.1170
−1.0	.1587	.1562	.1539	.1515	.1492	.1469	.1446	.1423	.1401	.1379
−0.9	.1841	.1814	.1788	.1762	.1736	.1711	.1685	.1660	.1635	.1611
−0.8	.2119	.2090	.2061	.2033	.2005	.1977	.1949	.1922	.1894	.1867
−0.7	.2420	.2389	.2358	.2327	.2296	.2266	.2236	.2206	.2177	.2148
−0.6	.2743	.2709	.2676	.2643	.2611	.2578	.2546	.2514	.2483	.2451
−0.5	.3085	.3050	.3015	.2981	.2946	.2912	.2877	.2843	.2810	.2776
−0.4	.3446	.3409	.3372	.3336	.3300	.3264	.3228	.3192	.3156	.3121
−0.3	.3821	.3783	.3745	.3707	.3669	.3632	.3594	.3557	.3520	.3483
−0.2	.4207	.4168	.4129	.4090	.4052	.4013	.3974	.3936	.3897	.3859
−0.1	.4602	.4562	.4522	.4483	.4443	.4404	.4364	.4325	.4286	.4247
−0.0	.5000	.4960	.4920	.4880	.4840	.4801	.4761	.4721	.4681	.4641

0.0	.5000	.5040	.5080	.5120	.5160	.5199	.5239	.5279	.5319	.5359
0.1	.5398	.5438	.5478	.5517	.5557	.5596	.5636	.5675	.5714	.5753
0.2	.5793	.5832	.5871	.5910	.5948	.5987	.6026	.6064	.6103	.6141
0.3	.6179	.6217	.6255	.6293	.6331	.6368	.6406	.6443	.6480	.6517
0.4	.6554	.6591	.6628	.6664	.6700	.6736	.6772	.6808	.6844	.6879
0.5	.6915	.6950	.6985	.7019	.7054	.7088	.7123	.7157	.7190	.7224
0.6	.7257	.7291	.7324	.7357	.7389	.7422	.7454	.7486	.7517	.7549
0.7	.7580	.7611	.7642	.7673	.7704	.7734	.7764	.7794	.7823	.7852
0.8	.7881	.7910	.7939	.7967	.7995	.8023	.8051	.8078	.8106	.8133
0.9	.8159	.8186	.8212	.8238	.8264	.8289	.8315	.8340	.8365	.8389
1.0	.8413	.8438	.8461	.8485	.8508	.8531	.8554	.8577	.8599	.8621
1.1	.8643	.8665	.8686	.8708	.8729	.8749	.8770	.8790	.8810	.8830
1.2	.8849	.8869	.8888	.8907	.8925	.8944	.8962	.8980	.8997	.9015
1.3	.9032	.9049	.9066	.9082	.9099	.9115	.9131	.9147	.9162	.9177
1.4	.9192	.9207	.9222	.9236	.9251	.9265	.9279	.9292	.9306	.9319
1.5	.9332	.9345	.9357	.9370	.9382	.9394	.9406	.9418	.9429	.9441
1.6	.9452	.9463	.9474	.9484	.9495	.9505	.9515	.9525	.9535	.9545
1.7	.9554	.9564	.9573	.9582	.9591	.9599	.9608	.9616	.9625	.9633
1.8	.9641	.9649	.9656	.9664	.9671	.9678	.9686	.9693	.9699	.9706
1.9	.9713	.9719	.9726	.9732	.9738	.9744	.9750	.9756	.9761	.9767
2.0	.9772	.9778	.9783	.9788	.9793	.9798	.9803	.9808	.9812	.9817
2.1	.9821	.9826	.9830	.9834	.9838	.9842	.9846	.9850	.9854	.9857
2.2	.9861	.9864	.9868	.9871	.9875	.9878	.9881	.9884	.9887	.9890
2.3	.9893	.9896	.9898	.9901	.9904	.9906	.9909	.9911	.9913	.9916
2.4	.9918	.9920	.9922	.9925	.9927	.9929	.9931	.9932	.9934	.9936
2.5	.9938	.9940	.9941	.9943	.9945	.9946	.9948	.9949	.9951	.9952
2.6	.9953	.9955	.9956	.9957	.9959	.9960	.9961	.9962	.9963	.9964
2.7	.9965	.9966	.9967	.9968	.9969	.9970	.9971	.9972	.9973	.9974
2.8	.9974	.9975	.9976	.9977	.9977	.9978	.9979	.9979	.9980	.9981
2.9	.9981	.9982	.9982	.9983	.9984	.9984	.9985	.9985	.9986	.9986
3.0	.9987	.9987	.9987	.9988	.9988	.9989	.9989	.9989	.9990	.9990
3.1	.9990	.9991	.9991	.9991	.9992	.9992	.9992	.9992	.9993	.9993
3.2	.9993	.9993	.9994	.9994	.9994	.9994	.9994	.9995	.9995	.9995
3.3	.9995	.9995	.9995	.9996	.9996	.9996	.9996	.9996	.9996	.9997
3.4	.9997	.9997	.9997	.9997	.9997	.9997	.9997	.9997	.9997	.9998

From Utts, Jessica M. and Robert F. Heckard. Mind on Statistics, Fourth Edition. 2012. Used with permission.

Try It! Finding Probabilities for *Z*

1. Find $P(Z \leq 1.22)$.

   Think about it: What is $P(Z < 1.22)$?

2. Find $P(Z > 1.22)$.

3. Find $P(-1.58 < Z < 2.24)$

4. What is the probability that a standard normal variable Z is within 2 standard deviations of the mean? That is, find $P(-2 \leq Z \leq 2)$.

5. What is $P(Z \leq 4.75)$?

In the Extreme (for $z > 0$)

z	3.09	3.72	4.26	4.75	5.20	5.61	6.00
Probability	.999	.9999	.99999	.999999	.9999999	.99999999	.999999999

6. What is $P(Z > 10.20)$?

7. What is the 90th percentile of the standard normal $N(0,1)$ distribution?

How to Solve General Normal Curve Problems

One way to find areas and thus probabilities for any general $N(\mu,\sigma)$ distribution involves the idea of standardization.

If the variable X has the $N(\mu,\sigma)$ distribution, then the standardized variable $Z = \dfrac{X-\mu}{\sigma}$ will have the $N(0,1)$ distribution.

We will see how this standardization idea works through our next example.

Try It! Scholastic Scores

A proposal before the Board of Education has specified certain designations for elementary schools depending on how their students do on the Evaluation of Scholastic Scoring. This test is given to all students in all public schools in grades 2 through 4. Schools that score in the top 20% are labeled excellent. Schools in the bottom 25% are labeled "in danger" and schools in the bottom 5% are designated as failing. Previous data suggests that scores on this test are approximately normal with a mean of 75 and a standard deviation of 5.

a. What is the probability that a randomly selected school will score below 70?

b. What is the score cut-off required for schools to be labeled excellent? Show all work.

Once you find the percentile on the standard normal scale, say *z*, then you can convert it to the corresponding percentile on the *X* scale by: $x = z\sigma + \mu$.

Notes:

(1) The normal distribution is a density for continuous random variables.

(2) The normal distribution is not the only continuous distribution (recall you worked with a family of uniform continuous distributions a few pages back).

(3) Computers and calculators often have the ability to find areas or percentiles for many density curves built right in to a function. You might be introduced to some of these in your lab sessions and are welcome to use them. But drawing a picture of what you are trying to find will be beneficial and serve as one way to show your work.

8.7 Approximating Binomial Distribution Probabilities

Recall Our Left-Handed Problem

In an earlier problem it was stated that about 10% of Americans are left-handed. Let X = the number of left-handed Americans in a random sample of 120 Americans (part c had us think about a sample size being 120 instead of 12).

Then X has an exact Binomial distribution with $n = 120$ and $p = 0.10$. The mean or expected number of left-handed Americans in the sample is $\mu = np = 120(0.10) = 12$. The standard deviation of X is $\sigma = \sqrt{np(1-p)} = \sqrt{120(0.10)(0.90)} = 3.29$.

Suppose we want to find the probability that a random sample of 120 will contain 20 or fewer left-handed Americans. Using the exact binomial distribution we would start with:

$$P(X \leq 20) = P(X = 0) + P(X = 1) + \cdots + P(X = 19) + P(X = 20)$$

This would not be too much fun to compute by hand since each of the probabilities for $X = 0$ up to $X = 20$ would be found using the binomial probability formula: $P(X = k) = \binom{n}{k} p^k (1-p)^{n-k}$.

But there is ***an easier way*** that will give us approximately the probability of having 20 or fewer. The easier way involves using a normal distribution. The normal distribution can be used to approximate probabilities for other types of random variables, one being binomial random variables when the sample size n is large.

Normal Approximation to the Binomial Distribution

If ***X* is a binomial** random variable based on n trials with success probability p, and ***n* is large**, then the random variable ***X* is also approximately** ...

Conditions:
The approximation works well when both np and $n(1-p)$ are at least 10.

Try It! Returning to our Left-Handed Problem

About 10% of Americans are left-handed. Let X = the number of left-handed Americans in a random sample of 120 Americans. Then X has an exact Binomial distribution with n = 120 and p = 0.10.

- **Mean** number of left-handed Americans in the sample $\mu = np = 120(0.10) = 12$.
- **Standard deviation** of $X = \sigma = \sqrt{np(1-p)} = \sqrt{120(0.10)(0.90)} = 3.29$.

a. We want to find the **probability that a random sample of 120 will contain 20 or fewer left-handed Americans**. Since $np = 120(0.10) = 12$ and $n(1-p) = 120(0.90) = 108$ are both at least 10, we can use the normal approximation for the distribution of X.

X has approximately a Normal distribution: $N($ ______, ______ $)$

$P(X \leq 20) =$

b. **How likely is it that more than 20% of the sample will be left-handed Americans**?

8.8 Sums, Differences, and Combinations of Random Variables

There are many instances where we want information about combinations of random variables. One type of combination of variables is a linear combination. Two primary linear combinations that arise are sums and differences.

Sum = $X + Y$ **Difference = $X - Y$**

The next two summary boxes present the rules for finding the mean and the variance (and thus standard deviation) of a sum and of a difference. We will see the results for a difference when we study learning about the difference between two proportions and about the difference between two means.

Rules for Means:

$$\text{Mean}(X + Y) = \text{Mean}(X) + \text{Mean}(Y)$$

$$\text{Mean}(X - Y) = \text{Mean}(X) - \text{Mean}(Y)$$

Rules for Variances (if X and Y are independent):

$$\text{Variance}(X + Y) = \text{Variance}(X) + \text{Variance}(Y)$$

$$\text{Variance}(X - Y) = \text{Variance}(X) + \text{Variance}(Y)$$

Think about it:
why is the variance of a difference found by taking the sum of the variances?

Additional Notes

A place to ... jot down questions you may have and ask during office hours, take a few extra notes, write out an extra practice problem or summary completed in lecture, create your own short summary about this chapter.

Stat 250 Gunderson Lecture Notes
Learning about a Population Proportion

Part 1: Introduction to Inference and Understanding Sampling Distributions
Chapter 9: Sections 1 – 4, SD Modules 0 and 1

To be a statistician is great!! You never have to be "absolutely sure" of something.
Being "reasonably certain" is enough!
-- *Pavel E. Guarisma, North Carolina State University* © FAIR USE

9.1 Parameters, Statistics, and Statistical Inference

Some distinctions to keep in mind ...

- **Population** versus **Sample**
- **Parameter** versus **Statistic**

  e.g. Population proportion p versus sample proportion $\hat{p}$
  e.g. Population mean μ versus sample mean $\overline{X}$

Since we hardly ever know the true population parameter value, we take a sample and use the sample statistic to estimate the parameter. When we do this, the sample statistic may not be equal to the population parameter, in fact, it could change every time we take a new sample. Will the observed sample statistic value be a reasonable estimate? If our sample is a ***RANDOM SAMPLE***, then we will be able to say something about the accuracy of the estimation process.

Statistical Inference:
the use of sample data to make judgments or decisions about populations.

The results from Chapter 9 will be used to make conclusions about population parameters based on sample statistics. The two most common statistical inference procedures are confidence interval estimation and hypothesis testing.

- **Confidence Interval Estimation**: A confidence interval is a range of values that the researcher is fairly confident will cover the true, unknown value of the population parameter. In other words, we use a confidence interval to estimate the value of a population parameter. We have already encountered the idea of a margin of error and using it to form a confidence interval for a population proportion in Chapter 5. We will explore confidence intervals further in Chapters 10 and 11.

- **Hypothesis Testing:** Hypothesis testing uses sample data to attempt to reject a hypothesis about the population. Usually researchers want to reject the notion that chance alone can explain the sample results. Hypothesis testing is applied to population parameters by specifying a null value for the parameter—a value that would indicate that nothing of interest is happening. Hypothesis testing proceeds by obtaining a sample, computing a sample statistic, and assessing how unlikely the sample statistic would be if the null parameter value were correct. In most cases, the researchers are trying to show that the null value is not correct. Achieving statistical significance is equivalent to rejecting the idea that the observed results are plausible if the null value is correct. We will explore hypothesis testing further in Chapters 12 and 13.

9.2 From Curiosity to Questions About Parameters

In Chapters 9 to 13, we study statistical inference methods for five parameters that are so commonly used that we call them "**the big five**." The five parameters all involve either **proportions** (for categorical data) or **means** (for quantitative data). Table 9.1 provides a summary of these big five scenarios and Section 9.2 is a great introduction to these five scenarios with examples of each. Read through this section – there are no computations, just shows how a question of interest can be translated into a statement about one of these big five parameters.

Table 9.1 Population Parameters and Sample Statistics for the Big Five Scenarios

Parameter Name and Description	Symbol for the Population Parameter	Symbol for the Sample Statistic
For Categorical Variables:		
One population proportion (or probability)	p	$\hat{p}$
Difference in two population proportions	$p_1 - p_2$	$\hat{p}_1 - \hat{p}_2$
For Quantitative Variables:		
One population mean	μ	$\overline{x}$
Population mean of paired differences (dependent)	μ_d	$\overline{d}$
Difference in two population means (independent)	$\mu_1 - \mu_2$	$\overline{x}_1 - \overline{x}_2$

From Utts, Jessica M. and Robert F. Heckard. Mind on Statistics, Fourth Edition. 2012. Used with permission.

This section ends with Table 9.2 that shows how Chapters 9 to 13 are organized. We will first focus on learning about a population proportion *p*. So we will cover each introductory module followed by module 1 for the one proportion scenario. We follow learning about one population proportion with that of learning about one population mean. Then turn to the remaining of the big five parameters.

9.3 SD Module 0: An Overview of Sampling Distributions

The value of a statistic from a **random sample** will vary from sample to sample. So **a statistic is a random variable** and **it will have a probability distribution**. This probability distribution is called the **sampling distribution** of the statistic.

> ***Definition:***
> The distribution of all possible values of a statistic for repeated samples of the same size from a population is called the **sampling distribution of the statistic.**

We will study the sampling distribution of various statistics, many of which will have approximately normal distributions. The general structure of the sampling distribution is the same for each of the five scenarios. The sampling distribution results, along with the ideas of probability and random sample, play a vital role in the inference methods that start with Chapter 10 and continue throughout the remainder of the course.

9.4 SD Module 1: Sampling Distributions for One Sample Proportion

Many responses of interest produce counts rather than measurements -- sex (male, female), political preference (republican, democrat), approve of new proposal (yes, no). We want to learn about a population proportion and we will do so using the information provided from a sample from the population.

Example: Do you work more than 40 hours per week?

A poll was conducted by The Heldrich Center for Workforce Development (at Rutgers University). A probability sample of 1000 workers resulted in 460 (for 46%) stating they work more than 40 hours per week.

Population =

Parameter =

Sample =

Statistic =

Can anyone say how close this observed sample proportion $\hat{p}$ is to the true population proportion p?

If we were to take another random sample of the same size $n = 1000$, would we get the same value for the sample proportion $\hat{p}$?

So what are the possible values for the sample proportion $\hat{p}$ if we took many random samples of the same size from this population? What would the distribution of the possible $\hat{p}$ values look like?

What is the sampling distribution of $\hat{p}$?

Aside: Can you Visualize It?

Consider taking your one random sample of size *n* and computing your one $\hat{p}$ value. (As in the previous example, our one $\hat{p}$ = 460/1000 = 0.46.) Suppose we did take another random sample of the same size, we would get another value of $\hat{p}$, say ___________. Now repeat that process over and over; taking one random sample after another; resulting in one $\hat{p}$ value after another.

Example picture showing the possible values when *n* = _______

0 0.1 0.2 0.3 0.4 0.5 0.6 0.7 0.8 0.9 1 $\hat{p}$ values

Observations:

How would things change if the sample size *n* were even larger, say *n* = _______? Suppose our first sample proportion turned out to be $\hat{p}$ = __________________ . Now imagine again repeating that process over and over; taking one random sample after another; resulting in many $\hat{p}$ possible values.

Example picture showing the possible values when *n* = _______

0 0.1 0.2 0.3 0.4 0.5 0.6 0.7 0.8 0.9 1 $\hat{p}$ values

Observations:

Let's take a closer look at the sample proportion $\hat{p}$. The sample proportion is found by taking the number of "successes" in the sample and dividing by the sample size. So the count variable X of the number of successes is directly related to the proportion of successes as $\hat{p} = \frac{X}{n}$.

In Section 8.7 we studied the distribution of our first statistic, the count statistic *X* (the number of successes in *n* independent trials when the probability of a success was *p*). We learned about its exact distribution called the **Binomial Distribution**. We also learned that when the sample size (or number of trials) *n* was large, the distribution of *X* could be approximated with a normal distribution.

Normal Approximation to the Binomial Distribution

If ***X* is a binomial** random variable based on *n* trials with success probability *p*, and ***n* is large**, then the random variable ***X* is also approximately** $N\left(np, \sqrt{np(1-p)}\right)$.

Conditions: The approximation works well when both *np* and $n(1-p)$ are at least 10.

So any probability question about a sample proportion could be converted to a probability question about a sample count, and vice-versa.

- If ***n is small***, we would need to convert the question to a count and use the binomial distribution to work it out.
- If ***n is large***, we could convert the question to a count and use the normal approximation for a count, OR use a related normal approximation for a sample proportion (for large *n*). The normal approximation for a sample proportion is called **"The Normal Curve Approximation Rule for Sample Proportions"** and is summarized below.

The **Normal Curve Approximation Rule for Sample Proportions** can be defined as follows:

Let p = population proportion of interest or binomial probability of success.
Let $\hat{p}$ = corresponding sample proportion or proportion of successes.

If numerous samples or repetitions of the same size n are taken, the distribution of possible values of $\hat{p}$ is approximately a normal curve distribution with

Mean $= p$

$$\text{Standard deviation} = \text{s.d.}(\hat{p}) = \sqrt{\frac{p(1-p)}{n}}$$

This approximate normal distribution is called the **sampling distribution of $\hat{p}$.**

From Utts, Jessica M. and Robert F. Heckard. Mind on Statistics, Fourth Edition. 2012. Used with permission.

Pages 324-325 provide a nice review of the conditions that need to be met for this Normal Curve Approximation Rule to apply.

The Stat 250 formula card summarizes this rule as follows:

> **Sampling Distribution of $\hat{p}$**
> If the sample size n is large enough
> (namely, $np \geq 10$ and $n(1-p) \geq 10$),
>
> then $\hat{p}$ is approximately $N\left(p, \sqrt{\frac{p(1-p)}{n}}\right)$.

Let's put this result to work in our next Try It! Problem.

Try It! Do Americans really vote when they say they do?

To answer this question, a telephone poll was taken two days after the November 8, 1994, election. From the 800 adults polled, 56% reported that they had voted. However, it was later reported in the press that, in fact, only 39% of American adults had voted. Suppose the 39% rate reported by the press is the correct population proportion. Also assume the responses of the 800 adults polled can be viewed as a random sample.

a. Describe and sketch the sampling distribution of $\hat{p}$ for a random sample of size n = 800 adults.

b. What is the approximate probability that a sample proportion who voted would be 56% or larger for a random sample of 800 adults?

c. Does it seem that the poll result of 56% simply reflects a sample that, by chance, voted with greater frequency than the general population?

More on the Standard Deviation of $\hat{p}$

The standard deviation of $\hat{p}$ is given by: s.d.($\hat{p}$) = $\sqrt{\dfrac{p(1-p)}{n}}$

This quantity would give us an idea about how far apart a sample proportion $\hat{p}$ and the true population proportion *p* are likely to be.

We can ***interpret*** **this standard deviation** as **approximately** the **average distance** of the possible $\hat{p}$ values (for repeated samples of the same size *n*) from the **true population proportion** *p*.

In practice when we take a random sample from a large population, we only know the sample proportion. We generally would not know the true population proportion *p*. So we could not compute the standard deviation of $\hat{p}$.

However we can use the sample proportion in the formula to have an estimate of the standard deviation, which is called the **standard error of** $\hat{p}$.

The standard error of $\hat{p}$ is given by: s.e.($\hat{p}$) = $\sqrt{\dfrac{\hat{p}(1-\hat{p})}{n}}$

This quantity is an *estimate* of the standard deviation of $\hat{p}$.

So we can ***interpret*** **this standard error** as ***estimating***, **approximately**, the **average distance** of the possible $\hat{p}$ values (for repeated samples of the same size *n*) from the **true population proportion** *p*.

Moreover, we can use this standard error to create a range of values that we are very confident will contain the true proportion *p*, namely, $\hat{p} \pm$ (a few)s.e.($\hat{p}$).

This is the basis for confidence interval for the true proportion *p*, discussed more in the next Chapter 10.

Try It! Love at first sight?

In a random sample of n = 500 adults, 300 stated they believe in love at first sight.

a. Estimate the population proportion of adults that believe in love at first sight.

b. Find the corresponding standard error of for the estimate in part a and use this standard error to provide an interval estimate for the population proportion p, with 95% confidence.

Additional Notes

A place to ... jot down questions you may have and ask during office hours, take a few extra notes, write out an extra practice problem or summary completed in lecture, create your own short summary about this chapter.

Stat 250 Gunderson Lecture Notes
Learning about a Population Proportion

Part 2: Estimating Proportions with Confidence
Chapter 10: Sections 1, 2, and 4, CI Modules 0 and 1

Big Idea of Confidence Intervals: Use sample data to estimate a population parameter.

10.1 CI Module 0: An Overview of Confidence Intervals

Recall some of the language and notation associated with the estimation process.

- **Population and Population Parameter**

- **Sample and Sample Statistic**
  (sample estimate or point estimate)

The **sample estimate** provides our best guess as to what is the value of the population parameter, but it is not 100% accurate.

The **value of the sample estimate will vary** from one sample to the next. The values often vary around the population parameter and the standard deviation give an idea about how far the sample estimates tend to be from the true population proportion on average.

The **standard error of the sample estimate** provides an idea of how far away it would tend to vary from the parameter value (on average).

The **general format** for a confidence interval estimate is given by:

Sample estimate ± (a few) standard errors

The "**few**" or number of standard errors we go out each way from the sample estimate will depend on how confident we want to be.

The "**how confident**" we want to be is referred to as the **confidence level**. This level reflects how confident we are in the *procedure*. Most of the intervals that are made will contain the truth about the population, but occasionally an interval will be produced that does not contain the true parameter value. Each interval either contains the population parameter or it doesn't. The confidence level is the percentage of the time we expect the *procedure* to produce an interval that does contain the population parameter.

10.2 CI Module 1:
Confidence Interval for a Population Proportion *p*

Goal: we want to learn about a population proportion $\hat{p}$. **How?** We take a random sample from the population and estimate *p* with the resulting sample proportion $\hat{p}$. In Chpt 9 we studied the sampling distribution of the statistic $\hat{p}$.

Sampling Distribution of $\hat{p}$

If the sample size *n* is large and $np \geq 10$ and $n(1-p) \geq 10$,

then $\hat{p}$ is approximately $N\left(p, \sqrt{\frac{p(1-p)}{n}}\right)$.

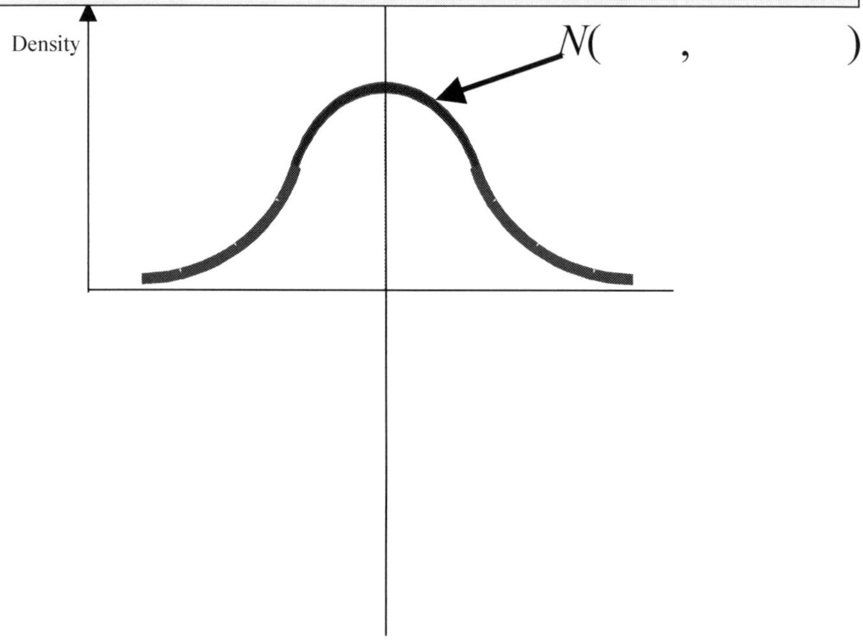

1. Consider the following interval or range of values and show it on the picture.
$$p \pm 2\sqrt{\frac{p(1-p)}{n}} \Rightarrow \left(p - 2\sqrt{\frac{p(1-p)}{n}}, p + 2\sqrt{\frac{p(1-p)}{n}}\right)$$
2. What is the probability that a (yet to be computed) sample proportion $\hat{p}$ will be in this interval (within 2 standard deviations from the true proportion *p*)?
3. Take a possible sample proportion $\hat{p}$ and consider the interval
$$\hat{p} \pm 2\sqrt{\frac{p(1-p)}{n}} \Rightarrow \left(\hat{p} - 2\sqrt{\frac{p(1-p)}{n}}, \hat{p} + 2\sqrt{\frac{p(1-p)}{n}}\right)$$
Show this range on the normal distribution picture above.
4. Did your first interval around your first $\hat{p}$ contain the true proportion *p*? Was it a 'good' interval? __________
5. Repeat steps 3 and 4 for other possible values of $\hat{p}$.

Big Idea:

- Consider all possible random samples of the same large size *n*.
- Each possible random sample provides a possible sample proportion value. (If we made a histogram of all of these possible $\hat{p}$ values it would look like the normal distribution on the previous page.)
- About 95% of the possible sample proportion $\hat{p}$ values will be in the interval $p \pm 2\sqrt{\frac{p(1-p)}{n}}$; and for each one of these sample proportion $\hat{p}$ values, the interval $\hat{p} \pm 2\sqrt{\frac{p(1-p)}{n}}$ will contain the population proportion *p*.
- Thus about 95% of the intervals $\hat{p} \pm 2\sqrt{\frac{p(1-p)}{n}}$ will contain the population proportion *p*. Also see Lesson 2 on pages 380-381 of the textbook for this development.

Thus, an initial 95% confidence interval for the true proportion *p* is given by:

$$\hat{p} \pm 2\sqrt{\frac{p(1-p)}{n}}$$

The Dilemma: When we take our one random sample, we can compute the sample proportion $\hat{p}$, but we can't construct the interval $\hat{p} \pm 2\sqrt{\frac{p(1-p)}{n}}$ because we don't know the value of *p*.

The Solution: Replace the value of *p* in the standard deviation with the estimate $\hat{p}$,

that is use $\sqrt{\frac{\hat{p}(1-\hat{p})}{n}}$ called ____________________.

An <u>approximate</u> 95% confidence interval (CI)
for the population proportion *p* is given by:

$$\hat{p} \pm 2\sqrt{\frac{\hat{p}(1-\hat{p})}{n}}$$

Note: The ± part of the interval $2\sqrt{\frac{\hat{p}(1-\hat{p})}{n}}$ is called the 95% <u>margin of error</u>.

Note: The **<u>approximate</u>** is due to the multiplier of '2' being used. We will learn about other multipliers, including the exact 95% multiplier value later in these notes.

Try It! Getting Along with Parents

In a Gallup Youth Survey n = 501 randomly selected American teenagers were asked about how well they get along with their parents. One survey result was that 54% of the sample said they get along "VERY WELL" with their parents.

a. The sample proportion was found to be 0.54. Give the standard error for the sample proportion and use it to complete the sentence that interprets the standard error in terms of an average distance.

We would estimate the average distance between the possible ______ values

(from repeated samples) and ______________________ to be about 0.022.

b. Compute a 95% confidence interval for the population proportion of teenagers that get along very well with their parents.

c. Fill in the blanks for the typical interpretation of the confidence interval in part b:

"Based on this sample, with 95% confidence, we would estimate that somewhere between _________ and ________ of *all* American teenagers think they get along very well with their parents."

d. Can we say the probability that the above (already observed) interval (____________ , ____________) contains the population proportion p is 0.95?

That is, can we say $P(\_\_\_\_\_\_\_\_\_ \leq p \leq \_\_\_\_\_\_\_\_\_) = 0.95?$

e. Can we say that 95% of the time the population proportion p will be in the interval computed in part b?

Just what does the 95% confidence level mean? Interpretation

The phrase **confidence level** is used to describe the likeliness or chance that a yet-to-be constructed interval will actually contain the true population value in the following sense (from page 374 of the text).

> **Interpreting the Confidence Level**
>
> - For a confidence interval, the **confidence level** is the probability that the procedure that is used to determine the interval will provide an interval that includes the population parameter.
> - The confidence level has a relative frequency probability interpretation: If we consider all possible randomly selected samples of the same size from a population, the *confidence level* is the fraction or percentage of those samples for which the confidence interval includes the population parameter.
>
> ***Note:*** It is commonplace to express the confidence level as a percentage.

From Utts, Jessica M. and Robert F. Heckard. Mind on Statistics, Fourth Edition. 2012. Used with permission.

The population proportion p is not a random quantity, it does not vary - once we have "looked" (computed) the actual interval, we cannot talk about probability or chance for this particular interval anymore. The 95% **confidence level applies to the procedure**, not to an individual interval; it applies "before you look" and not "after you look" at your data and compute your interval.

Try It! Getting Along with Parents

In the previous Try It! you computed a 95% confidence interval for the population proportion of teenagers that get along very well with their parents in part (b). This was based on a random sample of n = 501 American teenagers. You interpreted the interval in part (c). Write a sentence or two that ***interprets the confidence level***.

The interval we found was computed with a method which if repeated over and over ...

Try It! Completing a Graduate Degree

A researcher has taken a random sample of n = 100 recent college graduates and recorded whether or not the student completed their degree in 5 years or less. Based on these data, a 95% confidence interval for the population proportion of all college students that complete their degree in 5 years or less is computed to be (0.62, 0.80).

a. How many of the 100 sampled college graduates completed their degree in 5 years or less?

b. Which of the following statements gives a valid interpretation of this 95% confidence level? Circle all that are valid.

   i. There is about a 95% chance that the population proportion of students who have completed their degree in 5 years or less is between 0.62 and 0.80.

   ii. If the sampling procedure were repeated many times, then approximately 95% of the resulting confidence intervals would contain the population proportion of students who have completed their degree in 5 years or less.

   iii. The probability that the population proportion p falls between 0.62 and 0.80 is 0.95 for repeated samples of the same size from the same population.

What about that Multiplier of 2?

The exact multiplier of the standard error for a 95% confidence level would be 1.96, which was rounded to the value of 2. Where does the 1.96 come from? Use the standard normal distribution, the $N(0, 1)$ distribution at the right and Table A.1.

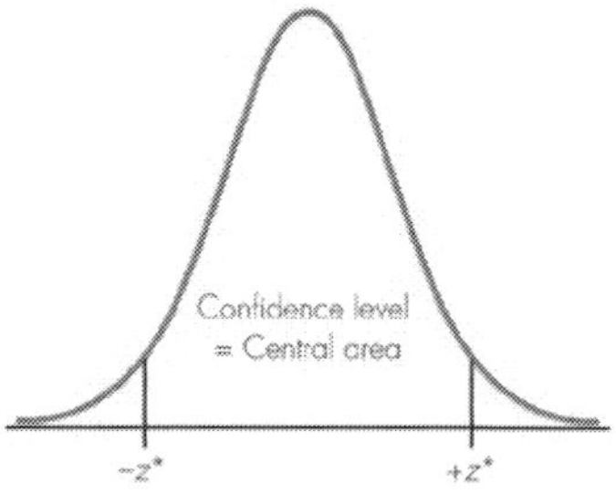

Researchers may not always want to use a 95% confidence level. Other common levels are 90%, 98% and 99%.

Using the same idea for confirming the value of 1.96, find the correct multiplier if the confidence level were 90%.

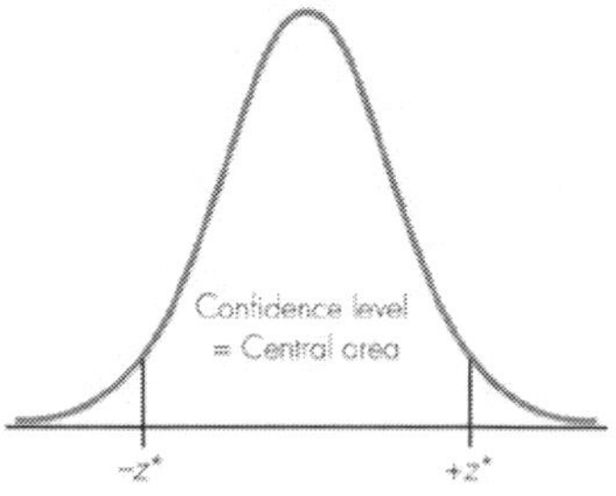

From Utts, Jessica M. and Robert F. Heckard. Mind on Statistics, Fourth Edition. 2012. Used with permission.

The generic expression for this multiplier when you are working with a standard normal distribution is given by z^*.

The table on the next page provides some of the multipliers for a population proportion confidence interval (page 378 of your text).

Table 10.1 Confidence Intervals for a Population Proportion

Confidence Level	Multiplier (z^*)	Confidence Interval
90	1.645 or 1.65.	$\hat{p} \pm 1.65$ standard errors
95	1.960, sometimes rounded to 2	$\hat{p} \pm 2$ standard errors
98	2.326 or 2.33	$\hat{p} \pm 2.33$ standard errors
99	2.576 or 2.58	$\hat{p} \pm 2.58$ standard errors

From Utts, Jessica M. and Robert F. Heckard. Mind on Statistics, Fourth Edition. 2012. Used with permission.

Now, the easiest way to find most multipliers is to actually look ahead a bit and make use of Table A.2. Look at the df row marked *Infinite* degrees of freedom and you will find the z* values for many common confidence levels. Check it out!

When the confidence level increases, the value of the multiplier increases. So the width of the confidence interval also increases. In order to be more confident in the procedure (have a procedure with a higher probability of producing an interval that will contain the population value, we have to sacrifice and have a wider interval. The formula for a confidence interval for a population proportion *p* is summarized next.

Confidence Interval for a Population Proportion *p*

$$\hat{p} \pm z^* \text{s.e.}(\hat{p})$$

where $\hat{p}$ is the sample proportion;

z^* is the appropriate multiplier value from the $N(0,1)$ distrib (Table A.2);

and s.e.($\hat{p}$) = $\sqrt{\dfrac{\hat{p}(1-\hat{p})}{n}}$ is the standard error of the sample proportion.

Conditions:

1. The sample is a randomly selected sample from the population. However, available data can be used to make inferences about a much larger group if the data can be considered to be representative with regard to the question(s) of interest (called the **Fundamental Rule for Using Data for Inference**).
2. The sample size *n* is large enough so that the normal curve approximation holds; i.e. $np \geq 10$ and $n(1-p) \geq 10$.

Try It! A 90% CI for *p*

A random sample of *n* = 501 American teenagers resulted in 54% stating they get along very well with their parents. The standard error for this estimate was found to be 2.2%. The 95% confidence interval for the population proportion of teenagers that get along very well with their parents went from 49.6% to 58.4%. The corresponding 90% confidence interval would go from 50.4% to 57.6%, which is indeed narrower (but still centered around the estimate of 54%).

The Conservative Approach (also see Lesson 3 on pages 383-385 of the textbook)

From the general form of the confidence interval, the margin of error is given as:

$$\text{Margin of error} = z^* \text{ s.e.}(\hat{p}) = z^* \sqrt{\frac{\hat{p}(1-\hat{p})}{n}}$$

For any fixed sample size *n*, this margin of error will be the largest when $\hat{p}$ = ½ = 0.5. (Think about what the function $\hat{p}(1-\hat{p})$ looks like.) So using ½ for $\hat{p}$ in the above margin of error expression we have:

$$z^*\sqrt{\frac{\hat{p}(1-\hat{p})}{n}} = z^*\sqrt{\frac{\frac{1}{2}(1-\frac{1}{2})}{n}} = \frac{z^*}{2\sqrt{n}}$$

By using this margin of error for computing a confidence interval, we are being *conservative*. The resulting interval may be a little wider than needed, but it will not err on being too narrow. This leads to a corresponding **conservative confidence interval** for a population proportion.

Conservative Confidence Interval for a Population Proportion *p*

$$\hat{p} \pm \frac{z^*}{2\sqrt{n}}$$

where $\hat{p}$ is the sample proportion;

z^* is the appropriate multiplier value from the $N(0,1)$ distribution (Table A.2).

In Chapter 3 the margin of error for a proportion was given as $1/\sqrt{n}$.

This is actually a **95% conservative margin of error**.

What happens to the conservative margin of error in the box above when you use z^* = 2 for 95% confidence?

Choosing a Sample Size for a Survey

The choice of a sample size is important in planning a survey. Often a sample size is selected (using the conservative approach) that such that it will produce a desired margin of error for a given level of confidence. Let's take a look at the conservative margin of error more closely.

$$\textbf{(Conservative) Margin of Error} = \frac{z^*}{2\sqrt{n}}$$

Solving this expression for the **sample size** n we have: $n = \left(\frac{z^*}{2m}\right)^2$

If this does is not a whole number, we would round up to the next largest integer.

Try It! Coke versus Pepsi

A poll was conducted in Canada to estimate *p*, the proportion of Canadian college students who prefer Coke over Pepsi. Based on the sampled results, a 95% conservative confidence interval for *p* was found to be (0.62, 0.70).

a. What is the margin of error for this interval?

b. What sample size would be necessary in order to get a conservative 95% confidence interval for *p* with a margin of error of 0.03 (that is, an interval with a width of 0.06)?

c. Suppose that the same poll was repeated in the United States (whose population is 10 times larger than Canada), but four times the number of people were interviewed. The resulting 95% conservative confidence interval for *p* will be:

   A. twice as wide as the Canadian interval

   B. 1/2 as wide as the Canadian interval

   C. 1/4 as wide as the Canadian interval

   D. 1/10 as wide as the Canadian interval

   E. the same width as the Canadian interval

10.4 Using Confidence Intervals to Guide Decisions

We will look a little here at two of the principles (from pages 389-390 of the textbook):

Principle 1.
A value not in a confidence interval can be rejected as a likely value of the population proportion. A value that is in a confidence interval is an "acceptable" possibility for the value of a population proportion.

Principle 3.
When the confidence intervals for proportions in two different populations do not overlap, it is reasonable to conclude that the two population proportions are different.

However, if the intervals do overlap, no conclusion can be made.
A confidence interval for the difference in two proportions should be constructed to determine if 0 is a plausible value. We will discuss this more on this later.

Try It! Coke versus Pepsi

Recall the poll conducted in Canada to estimate p, the proportion of Canadian college students who prefer Coke over Pepsi.

Based on the sampled results, a 95% conservative confidence interval for p was found to be (0.62, 0.70).

Do you think it is reasonable to conclude that a majority of Canadian college students prefer Coke over Pepsi? Explain.

Additional Notes

A place to ... jot down questions you may have and ask during office hours, take a few extra notes, write out an extra practice problem or summary completed in lecture, create your own short summary about this chapter.

Stat 250 Gunderson Lecture Notes
Learning about a Population Proportion

Part 3: Testing about a Population Proportion
Chapter 12: Sections 1, 2, and 4, HT Modules 0 and 1

We make decisions in the dark of data. -- *Stu Hunter*

In Chapter 10 we examined statistical methods for estimating the population proportion based on the sample proportion using a confidence interval estimate. In Chapter 12 we turn to methods for testing theories about the population proportion. The hypothesis testing method uses data from a sample to judge whether or not a statement about a population is reasonable or not. We want to test theories about a population proportion and we will do so using the information provided from a sample from the population.

The **Basic Steps in Any Hypothesis Test** are given next:

Step 1: Determine the null and alternative hypotheses.
Step 2: Verify necessary data conditions, and if met, summarize the data into an appropriate test statistic.
Step 3: Assuming the null hypothesis is true, find the *p*-value.
Step 4: Decide whether or not the result is statistically significant based on the *p*-value.
Step 5: Report the conclusion in the context of the situation.

12.1 HT Module 0: An Overview of Hypothesis Testing

Formulating Hypothesis Statements

Many questions in research can be expressed as which of two statements might be correct for a population. These two statements are called the **null** and the **alternative hypotheses**.

> The **null hypothesis,** represented by the symbol $\mathbf{H_0}$, is a statement that there is nothing happening. The specific null hypothesis varies from problem to problem, but generally it can be thought of as the status quo, or no relationship, or no difference. In most situations, the researcher hopes to disprove or reject the null hypothesis.
>
> The **alternative hypothesis,** represented by the symbol $\mathbf{H_a}$, is a statement that something is happening. In most situations, this hypothesis is what the researcher hopes to prove. It may be a statement that the assumed status quo is false, or that there is a relationship, or that there is a difference.

From Utts, Jessica M. and Robert F. Heckard. Mind on Statistics, Fourth Edition. 2012. Used with permission.

The null and alternative hypotheses are statements about a population parameter (not about the results in the sample). There will be a ***"direction of extreme"*** that is indicated by the alternative hypothesis. To see these ideas, let's try writing out some hypotheses to be put to the test.

Try It! Stating the Hypotheses and defining the parameter of interest

1. About 10% of the human population is left-handed. Suppose that a researcher speculates that artists are more likely to be left-handed than are other people in the general population.

   H_0: let ________ = ____________________________

   H_a: parameter = written description

   Direction:

2. Suppose that a pharmaceutical company wants to be able to claim that for its newest medication the proportion of patients who experience side effects is less than 20%.

   H_0: let ________ = ____________________________

   H_a: parameter = written description

   Direction:

3. The US Census reports that 48% of households have no children. A random sample of 500 households will be taken to assess if the population proportion has changed from the Census value of 0.48.

   H_0: let ________ = ____________________________

   H_a: parameter = written description

   Direction:

Notes:

1. When the alternative hypothesis specifies a single direction, the test is called a **one-sided or one-tailed hypothesis test**. In practice, most hypothesis tests are one-sided tests because investigators usually have a particular direction in mind when they consider a question.
2. When the alternative hypothesis includes values in both direction from a specific standard, the test is called a **two-sided or two-tailed hypothesis test.**
3. A generic null hypothesis could be expressed as **H_0: population parameter = null value,** where the null value is the specific number the parameter equals if the null hypothesis is true. In all of the examples above, the population parameter is *p*, the population proportion. In Example 1 above the null value is 10% or 0.10.

The Logic of Hypothesis Testing: What if the Null is True?

The logic in hypothesis testing is similar to that of a jury trial ...

H_0: The defendant is ________________ **H_a:** The defendant is ________________

We assume that the null hypothesis is true until the sample data conclusively demonstrate otherwise. We assess whether or not the observed data are consistent with the null hypothesis (allowing reasonable variability). If the data are "unlikely" when the null hypothesis is true, we would reject the null hypothesis and support the alternative theory.

The Big Question we ask: If the null hypothesis is true about the population, what is the probability of observing sample data like that observed (or more extreme)?

Reaching Conclusions about the Two Hypotheses

We will be deciding between the two hypotheses using data. The data is assumed to be a **random sample** from the population under study.

The data will be summarized via a ________________________________. In many cases the test statistic is a *standardized statistic* that measures the distance between the sample statistic and the null value in standard error units.

$$\textbf{Test Statistic} = \frac{\textbf{Sample Statistic} - \textbf{Null Value}}{\textbf{(Null) Standard Error}}$$

In fact, our first test statistic will be a z-score and we are already familiar with what makes a z-value unusual or large.

With the test statistic computed, we quantify the compatibility of the result with the null hypothesis with a probability value called the *p*-value.

> The ***p*-value** is computed by assuming the null hypothesis is true and then determining the probability of a result as extreme (or more extreme) as the observed test statistic in the direction of the alternative hypothesis.

Notes:

(1) The *p*-value is a probability, so it must be between 0 and 1. It is really a conditional probability – the probability of seeing a test statistic as extreme or more extreme than observed **given (or conditional on)** the null hypothesis is true.

(2) The *p*-value is **not** the probability that the null hypothesis is true.

The ______________________ the *p*-value,
the stronger the evidence is **AGAINST H_0** (and in favor of H_a).

Common Convention: Reject H_0 if the *p*-value is ______________________________.

This borderline value is called the ______________________ and denoted by _____.

When the *p*-value is $\leq \alpha$, we say the result is ______________________________.

Common levels of significance are: ____________________________________

Two Possible Results:

- The **<u>*p*-value is $\leq \alpha$</u>**,
  so we **<u>reject H_0</u>**
  and say the **<u>results are statistically significant</u> <u>at the level α</u>**.
  We would then write a real-world conclusion to explain what 'rejecting H_0' translates to in the context of the problem at hand.

- The **<u>*p*-value is > α</u>**,
  so we **<u>fail to reject H_0</u>**
  and say the **<u>results are not statistically significant at the level α</u>**.
  We would then write a real-world conclusion to explain what 'failing to reject H_0' translates to in the context of the problem at hand.

Be careful: we say "fail to reject H_0" and not "accept H_0" because the data do not prove the null hypothesis is true, rather the data were not convincing enough to support the alternative hypothesis. Pages 457-458 provide a nice example.

12.2 HT Module 1: Testing Hypotheses About a Population Proportion

In the context of testing about the value of a population proportion p, the possible hypotheses statements are:

1. H_0: **versus** H_a:

2. H_0: **versus** H_a:

3. H_0: **versus** H_a:

Where does p_0 come from?

Sometimes the null hypothesis is written as $H_0: p = p_0$ as we compute the p-value assuming the null hypothesis is true, that is, we take the population proportion to be the null value p_0.

The sample data will provide us with an estimate of the population proportion p, namely the sample proportion $\hat{p}$. For a large sample size, the distribution for the sample proportion will be ...

If we have a normal distribution for a variable, then we can **standardize** that variable to compute probabilities, as long as you have the mean and standard deviation for that statistic. In testing, we assume that the null hypothesis is true, that the population proportion $p = p_0$. So the standardized z-statistic for a sample proportion in testing is:

$z =$

If the null hypothesis is true,

this z-test statistic will have approximately a ________________________.

The standard normal distribution will be used to compute the p-value for the test.

Try It! Left-handed Artists

About 10% of the human population is left-handed. Suppose that a researcher speculates that artists are more likely to be left-handed than are other people in the general population. The researcher surveys a random sample of 150 artists and finds that 18 of them are left-handed. Perform the test using a 5% significance level.

Step 1: Determine the null and alternative hypotheses.

H_0: ____________________ H_a: ____________________

where the parameter _____ represents ____________________________

____

Note: The direction of extreme is

Step 2: Verify necessary data conditions, and if met, summarize the data into an appropriate test statistic.

- The data are assumed to be a random sample.
- Check if $np_0 \geq 10$ and $n(1 - p_0) \geq 10$.

Observed test statistic: $z = \dfrac{\hat{p} - p_0}{\sqrt{\dfrac{p_0(1-p_0)}{n}}}$

Step 3: Assuming the null hypothesis is true, find the *p*-value.

The *p*-value is the probability of getting a test statistic as extreme or more extreme than the observed test statistic value, assuming the null hypothesis is true. Since we have a **one-sided test to the right**, toward the larger values ...

p-value = probability of getting a *z* test statistic as large or larger than observed, assuming the null hypothesis is true.

=

Step 4: Decide if the result is statistically significant based on the *p*-value.

Step 5: Report the conclusion in the context of the situation.

Aside:
The researcher chooses the level of significance α **before** the study is conducted.

In our Left-Handed Artists example we had H_0: $p = 0.10$ versus H_a: $p > 0.10$.

If $\hat{p} =$ 0.08, we would certainly not reject H_0

If $\hat{p} =$ 0.12, our z=0.82,
p-value=0.206 and our decision was to fail to reject H_0

If $\hat{p} =$ 0.16, our z=2.45, *p*-value=0.007,
so our decision would be ... ______________

If $\hat{p} =$ 0.19, our z=3.7, *p*-value=0.0000,
so our decision would be ... ______________

So, if $\hat{p} = 0.29$, our z=7.8, *p*-value nearly 0,
so our decision would be ... ______________

Selecting the level of significance is like drawing a line in the sand – separating when you will reject H_0 and when there the evidence would be strong enough to reject H_0.

This first test was a one-sided test to the right. How is the *p*-value found for the other directions of extreme? The table below (page 465 of the text) provides a nice summary.

Table 12.1 Alternative Hypothesis Regions and *p*-Value Areas

Statement of H_a		*p*-Value Area	Normal Curve Region
$p < p_0$	(less than)	Area to the left of z (even if $z > 0$)	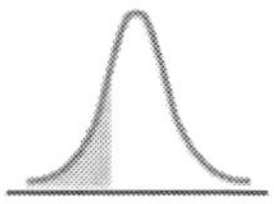
$p > p_0$	(greater than)	Area to the right of z (even if $z < 0$)	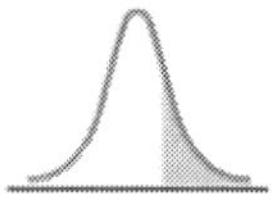
$p \neq p_0$	(not equal)	$2 \times$ area to the right of $\|z\|$	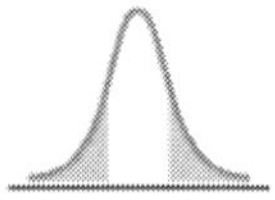

From Utts, Jessica M. and Robert F. Heckard. Mind on Statistics, Fourth Edition. 2012. Used with permission.

Try It! Households without Children

The US Census reports that 48% of households have no children. A random sample of 500 households was taken to assess if the population proportion has changed from the Census value of 0.48. Of the 500 households, 220 had no children. Use a 10% significance level.

Step 1: Determine the null and alternative hypotheses.

H_0: ***p = 0.48*** H_a: ***p ≠ 0.48***

where the parameter *p* represents the population proportion of all households *today* that have no children.

Note: The direction of extreme is **two-sided**.

Step 2: Verify necessary data conditions, and if met, summarize the data into an appropriate test statistic.

- The data are assumed to be a random sample.
- Check if $np_0 \geq 10$ and $n(1 - p_0) \geq 10$.

Observed test statistic: $z = \dfrac{\hat{p} - p_0}{\sqrt{\dfrac{p_0(1-p_0)}{n}}}$

Step 3: Assuming the null hypothesis is true, find the *p*-value.

The *p*-value is the probability of getting a test statistic as extreme or more extreme than the observed test statistic value, assuming the null hypothesis is true. Since we have a **two-sided test**, both large and small values are "extreme".

Sketch the area that corresponds to the *p*-value·

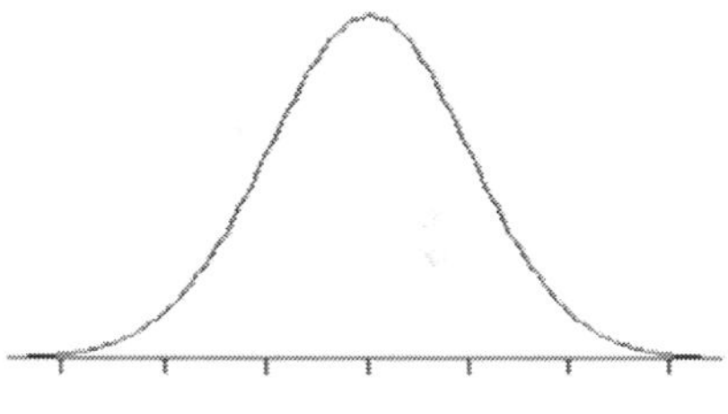

Compute the *p*-value:

Step 4: Decide if the result is statistically significant based on the *p*-value.

Step 5: Report the conclusion in the context of the situation.

What if *n* is small?

Goal: we still want to learn about a population proportion *p*.
We take a random sample of size *n* where *n* is small
(i.e. $np_0 < 10$ or $n(1 - p_0) < 10$).

If the sample size is small, we have to go back to the exact distribution for a count *X*, called the binomial distribution.

> If X has the ***binomial distribution*** *Bin*(*n*, *p*), then
>
> $$P(X = k) = \binom{n}{k} p^k (1-p)^{n-k} \quad \text{for } k = 0,1,2,\ldots,n \qquad \text{where } \binom{n}{k} = \frac{n!}{k!(n-k)!}$$
>
> and the mean of $X = \mu = np$
>
> and the standard deviation of $X = \sqrt{np(1-p)}$

We will use the binomial probability formula for computing the exact *p*-value.

Testing Hypotheses about a Population Proportion *p* when *n* is small

With a small sample size, we will do a Binomial test.

> ***Small-Sample Binomial Test for the population proportion p***
>
> To test the hypothesis H_0: $p = p_0$ we compute the count test statistic
>
> *X* = the number of successes in the sample of size *n*
> which has the *Bin*(*n*, p_0) distribution when H_0 is true.
>
> This *Bin*(*n*, p_0), distribution is used to compute the *p*-value for the test.

Try It! New Treatment

A group of 10 subjects with a disease are treated with a new treatment. Of the 10 subjects, 9 showed improvement. Test the claim that "a majority" of people using this treatment improved using a 5% significance level. Let p be the true population proportion of people who improve with this treatment.

State the hypotheses: H_0: ____________________ H_a: __________________

The observed test statistic value is just ____________________

p-value =

At the 5% significance level, we would ______________________ and conclude:

Let's revisit the flow chart for working on problems that deal with a population proportion.

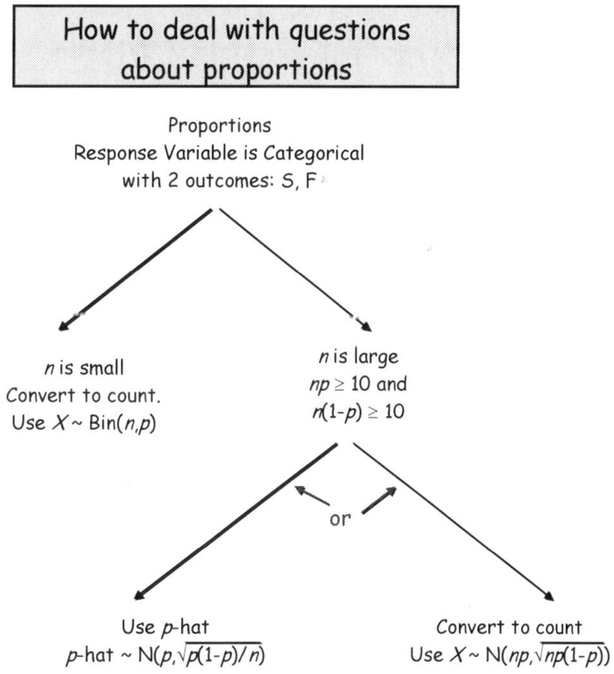

When the sample size is large we can use the large sample normal approximation for computing probabilities about a sample proportion, for testing hypotheses about a population proportion (based on the resulting sample proportion), and for computing a confidence interval estimate for the value of a population proportion (again using the sample proportion as the point estimate).

When the sample size is small we use the binomial distribution to compute probabilities about a sample proportion or to test hypotheses about a population proportion (based on the resulting sample count of successes). We did not discuss the small sample confidence interval for a population proportion using the binomial distribution.

The last section of Chapter 12 is important to read in full, including the examples. The main highlights are provided next.

12.4 Sample Size, Statistical Significance, and Practical Importance

The size of the sample affects our ability to make firm conclusions based on that sample. With a small sample, we may not be able to conclude anything. With large samples, we are more likely to find statistically significant results even though the actual size of the effect is very small and perhaps unimportant.

> **Cautions About Sample Size and Statistical Significance**
>
> - When there is a small to moderate effect in the population, a small sample has little chance of providing statistically significant support for the alternative hypothesis.
> - With a large sample, even a small and unimportant effect in the population may lead to a conclusion of statistical significance.

From Utts, Jessica M. and Robert F. Heckard. Mind on Statistics, Fourth Edition. 2012. Used with permission.

The phrase **statistically significant** only means that the data are strong enough to reject the null hypothesis. The *p*-value tells us about the statistical significance of the effect, but it does not tell us about the size of the effect. Example 12.18 is a good one to read and here is one below to check out.

Consider testing H_0: $p = 0.5$ versus H_a: $p > 0.5$ at $\alpha = 0.05$.

Warning #1: Small samples make it very difficult to demonstrate much of anything.

Case 1: 52 successes in a sample of size $n = 100$ → $\hat{p}$ = **0.52**

- Test Statistic: z = (0.52 - 0.50) / √ [0.5(1-0.5)/**100**] = 0.4
- *p*-value = $P(Z \geq 0.4) = 0.34$. So we would fail to reject H_0.
- An increase of only 0.02 beyond 0.50 seems inconsequential (not significant).

Case 2: 520 successes in a sample of size $n = 1000$ → $\hat{p}$ = **0.52**

- Test Statistic: z = (0.52 - 0.50) / √ [0.5(1-0.5)/**1000**] = 1.26
- *p*-value = $P(Z \geq 1.26) = 0.104$. So we would again fail to reject H_0.
- Here an increase of 0.02 beyond 0.50 is approaching significance.

Case 3: 5200 successes in a sample of size $n =$ 10,000 → $\hat{p}$ = **0.52**

- Test Statistic: z = (0.52 - 0.50) / √ [0.5(1-0.5)/**10,000**] = 4.0
- *p*-value = $P(Z \geq 4) = 0.00003$. So we would certainly reject H_0.
- Here an increase of 0.02 beyond 0.50 is very significant!

Warning #2: A huge sample size can make a practically **un**important difference statistically significant.

The Challenge: To determine an appropriate sample size so that findings that are *practically important* become *statistically significant*.

What Can Go Wrong: Two Types of Errors

We have been discussing a statistical technique for making a decision between two competing theories about a population. We base the decision on the results of a random sample from that population.

There is the possibility of making a mistake. In fact there are two types of error that we could make in hypothesis testing.

Two types of errors:

- Type 1 error = rejecting H_0 when H_0 is true
- Type 2 error = failing to reject H_0 when H_a is true

In statistics we have notation to represent the probabilities that a testing procedure will make these two types of errors.

P(Type 1 error) =

P(Type 2 error) =

There is another probability that is of interest to researchers – if there really is something going on (if the alternative theory is really true), what is the probability that we will be able to detect it (be able to reject H_0)? This probability is called the **power of the test** and is related to the probability of making a Type 2 error.

Power = P(rejecting H_0 when H_a is true)
= 1 – P(failing to reject H_0 when H_a is true)
= 1 – P(Type 2 error)
= $1 - \beta$.

So we can think of power as the probability of advocating the "new theory" given the "new theory" is true. Researchers are generally interested in having a test with high power.

One dilemma is that the best way to increase power is to increase sample size *n* (see comments below) and that can be expensive.

Comments:

1. In practice we want to protect the status quo
   so we are most concerned with _____.

2. Most tests we describe have the ...

3. Generally, for a fixed sample size *n*, ...

4. Ideally we want the probabilities of making a mistake to be small, we want the power of the test to be large. However, these probabilities are properties of the procedure (the proportion of times the mistake would occur if the procedure were repeated many times) and not applicable to the decision once it is made.

5. Some factors that influence the power of the test ...
   - Sample size: larger sample size leads to higher power.
   - Significance level: larger α leads to higher power.
   - Actual parameter value: a true value that falls further from the null value (in the direction of the alternative hypothesis) leads to higher power (however this is not something that the researcher can control or change).

Simple Example

H_0: Basket has 9 Red and 1 White
H_a: Basket has 4 Red and 6 White

Data: 1 ball selected at random from the basket.

What is the most reasonable Decision Rule?

Reject the null hypothesis if the ball is ____________________

With this rule, what are the chances of making a mistake?

P(Type 1 error) =

P(Type 2 error) =

What is the **power** of the test?

Suppose a ball is now selected from the basket and it is observed and found to be WHITE. What is the decision?

You just made a decision, could a mistake have been made?
If so, which type?

What is the probability that this type of mistake *was made*?

Note: The Decision Rule stated in the simple example resembles the **rejection region approach to hypothesis testing**. This is discussed on pages 473-475 of the textbook. We will focus primarily on the *p*-value approach that is used in reporting results in journals.

Additional Notes

A place to ... jot down questions you may have and ask during office hours, take a few extra notes, write out an extra practice problem or summary completed in lecture, create your own short summary about this chapter.

Stat 250 Gunderson Lecture Notes
Learning about the Difference in Population Proportions
Part 1: Distribution for a Difference in Sample Proportions
Chapter 9: Section 5 SD Module 2

The Independent Samples Scenario

Two samples are said to be **independent samples** when the measurements in one sample are not related to the measurements in the other sample. Independent samples are generated in a variety of ways. Some common ways:

- **Random samples are taken separately from two populations** and the same response variable is recorded for each individual.
- **One random sample** is taken and a variable is recorded for each individual, but then **units are categorized as belonging to one population or another**, e.g. male/female.
- **Participants are randomly assigned to one of two treatment conditions**, such as diet or exercise, and the same response variable, such as weight loss, is recorded for each individual unit.

If the **response variable is categorical**, a researcher might compare two independent groups by looking at the **difference between the two proportions**. Here is an example of an independent samples scenario for proportions from page 318 of your text.

> **Situation 2. Estimating the Difference Between Two Populations with Regard to the Proportion Falling into a Category of a Categorical Variable**
>
> - *Example research questions:*
>
>   How much difference is there between the proportions that would quit smoking if taking the antidepressant buproprion (Zyban) versus if wearing a nicotine patch?
>
>   How much difference is there between men who snore and men who don't snore with regard to the proportion who have heart disease?
> - *Population parameter:* $p_1 - p_2$, where p_1 and p_2 represent the proportions in populations 1 and 2, respectively
> - *Sample estimate:* $\hat{p}_1 - \hat{p}_2$, the difference between the two sample proportions

There are usually two questions of interest about a difference in two population proportions. First, we want to estimate the value of the difference. Second, often we want to test the hypothesis that the difference is 0, which would indicate that the two proportions are equal. In either case, we will need to know about the sampling distribution for the difference in two sample proportions (from independent samples).

9.5 SD Module 2: Sampling Distribution for the Difference in Two Sample Proportions

Example: Driving Safely

Question of interest: How much of a difference is there between men and women with regard to the proportion who have driven a car when they had too much alcohol to drive safely?

Study: Time magazine reported the results of a poll of adult Americans. One question asked was: **"Have you ever driven a car when you probably had too much alcohol to drive safely?"**

Let p_1 be the **population proportion** of **men** who would respond yes.
Let p_2 be the **population proportion** of **women** who would respond yes.

We want to learn about p_1 and p_2 and how they compare to each other. We could estimate the difference $p_1 - p_2$ with the corresponding difference in the sample proportions $\hat{p}_1 - \hat{p}_2$.

Will it be a good estimate? How close can we expect the difference in sample proportions to be to the true difference in population proportions (on average)?

Imagine repeating the study many times, each time taking two independent random samples of sizes n_1 and n_2, and computing the value of $\hat{p}_1 - \hat{p}_2$. What kind of values could you get for $\hat{p}_1 - \hat{p}_2$? What would the distribution of the possible $\hat{p}_1 - \hat{p}_2$ values look like? What can we say about the **distribution of the** difference in two sample proportions?

Using results from Section 8.8 (differences of independent random variables) and Section 9.4 (sampling distribution for a sample proportion), the sampling distribution of the difference in two sample proportions $\hat{p}_1 - \hat{p}_2$ can be determined.

From Section 8.8 we learned that the mean of the difference is just the difference in the two means, and that the variance of the difference of two independent random variables is the sum of the variances.

From Section 9.4 we saw the standard deviation of a sample proportion is $\sqrt{\frac{p(1-p)}{n}}$.

We will apply these ideas to our newest parameter of interest, the difference in two sample proportions $\hat{p}_1 - \hat{p}_2$.

Sampling Distribution of the Difference in Two (Independent) Sample Proportions

If the two sample proportions are based on independent random samples from the two populations and if all of the quantities $n_1\hat{p}_1$, $n_1(1-\hat{p}_1)$, $n_2\hat{p}_2$, and $n_2(1-\hat{p}_2)$ are at least 10, then $\hat{p}_1 - \hat{p}_2$ is (approximately)

Since the population proportions of p_1 and p_2 are not known, we will use the data to compute the standard error of the difference in sample proportions.

Standard Error of the Difference in Sample Proportions

$$\text{s.e.}(\hat{p}_1 - \hat{p}_2) = \sqrt{\frac{\hat{p}_1(1-\hat{p}_1)}{n_1} + \frac{\hat{p}_2(1-\hat{p}_2)}{n_2}}$$

The standard error of $\hat{p}_1 - \hat{p}_2$ estimates, roughly, the average distance of the possible $\hat{p}_1 - \hat{p}_2$ values from $p_1 - p_2$. The possible $\hat{p}_1 - \hat{p}_2$ values result from considering all possible independent random samples of the same sizes from the same two populations.

Moreover, we can use this standard error to produce a range of values that we can be quite confident will contain the difference in the population proportions $p_1 - p_2$:

$$\hat{p}_1 - \hat{p}_2 \pm \text{(a few)s.e.}(\hat{p}_1 - \hat{p}_2).$$

This is the basis for confidence interval for the difference in population proportions discussed in Chapter 10.

We will construct a null standard error of the difference in the sample proportions and use it to compute a standardized test statistic for testing hypotheses about the difference in the population proportions $p_1 - p_2$:

$$\frac{\textbf{Sample statistic – Null value}}{\textbf{(Null) standard error}}.$$

This is the basis for the hypothesis testing about the difference in population proportions covered in Chapter 12.

Additional Notes

A place to … jot down questions you may have and ask during office hours, take a few extra notes, write out an extra practice problem or summary completed in lecture, create your own short summary about this chapter.

Stat 250 Gunderson Lecture Notes
Learning about the Difference in Population Proportions

Part 2: Confidence Interval for a Difference in Population Proportions Chapter 10: Section 3 CI Module 2

10.3 CI Module 2: Confidence Interval for the Difference in Two Population Proportions

- We have **two populations** from which independent samples are available, (or one population for which two groups formed using a categorical variable).
- The response variable is also **categorical** and we are interested in comparing the proportions for the two populations.

Let p_1 be the population proportion for the first population.
Let p_2 be the population proportion for the second population.

Parameter: the difference in the population proportions $p_1 - p_2$.
Sample estimate: the difference in the sample proportions $\hat{p}_1 - \hat{p}_2$.
Standard error: $\text{s.e.}(\hat{p}_1 - \hat{p}_2) = \sqrt{\frac{\hat{p}_1(1-\hat{p}_1)}{n_1} + \frac{\hat{p}_2(1-\hat{p}_2)}{n_2}}$

So we have our estimate of the difference in the two population proportions, namely $\hat{p}_1 - \hat{p}_2$, and we have its standard error. To make our confidence interval, we need to know the multiplier.

Sample Estimate $\pm$ Multiplier x Standard error

As in the case for estimating one population proportion, we assume the sample sizes are sufficiently large so the multiplier will be a z^* value found from using the standard normal distribution.

Two Independent-Samples *z* Confidence Interval for p_1 - p_2

$$(\hat{p}_1 - \hat{p}_2) \pm z^* \text{s.e.}(\hat{p}_1 - \hat{p}_2)$$

where $\text{s.e.}(\hat{p}_1 - \hat{p}_2) = \sqrt{\frac{\hat{p}_1(1-\hat{p}_1)}{n_1} + \frac{\hat{p}_2(1-\hat{p}_2)}{n_2}}$

and z^* is the appropriate value for a standard normal distribution (found from Table A.2 with infinite degrees of freedom).

This interval requires that the sample proportions are based on independent random samples from the two populations. Also, all of the quantities $n_1\hat{p}_1$, $n_1(1-\hat{p}_1)$, $n_2\hat{p}_2$, and $n_2(1-\hat{p}_2)$ be preferably at least 10.

Try It! Driving Safely: Comparing DWI rates

Question: How much of a difference is there between men and women with regard to the proportion who have driven a car when they had too much alcohol to drive safely?

Study: Time magazine reported the results of a poll of adult Americans. One question asked was: **"Have you ever driven a car when you probably had too much alcohol to drive safely?"**

Results:

- Of the 300 men, 189 (63%) said yes, 108 (36%) said no; the remaining 3 men weren't sure.
- Of the 300 women, 87 (29%) said yes, 210 (70%) said no; the remaining 3 women weren't sure.

Compute a 95% confidence interval for the difference in the population proportion of men versus women who would say yes to this question.
Let 1 = men and 2 = women.
Note: the researchers wish to consider the responses as either 'yes' or 'not yes', thus keeping the sample sizes for both groups at 300.

Complete the sentence to interpret this interval.

With 95% confidence we estimate the difference in the population proportion of men versus women who have driven after having too much to drink to be somewhere between ____________ and ____________.

What value do you notice is *not* in this interval? _________

Does there appear to be a significant difference between the population rates for men versus women? **Yes** **No**

Here is the summary of the Confidence intervals for the big 5 parameters covered in Chapters 10 and 11. We have now covered the one and two population proportion scenarios.

Population Proportion	**Two Population Proportions**	**Population Mean**
Parameter p	**Parameter** $p_1 - p_2$	**Parameter** μ
Statistic $\hat{p}$	**Statistic** $\hat{p}_1 - \hat{p}_2$	**Statistic** $\bar{x}$
Standard Error $\text{s.e.}(\hat{p}) = \sqrt{\frac{\hat{p}(1-\hat{p})}{n}}$	**Standard Error** $\text{s.e.}(\hat{p}_1 - \hat{p}_2) = \sqrt{\frac{\hat{p}_1(1-\hat{p}_1)}{n_1} + \frac{\hat{p}_2(1-\hat{p}_2)}{n_2}}$	**Standard Error** $\text{s.e.}(\bar{x}) = \frac{s}{\sqrt{n}}$
Confidence Interval $\hat{p} \pm z^* \text{s.e.}(\hat{p})$ **Conservative Conf. Interval** $\hat{p} \pm \frac{z^*}{2\sqrt{n}}$ **Sample Size** $n = \left(\frac{z^*}{2m}\right)^2$	**Confidence Interval** $(\hat{p}_1 - \hat{p}_2) \pm z^* \text{s.e.}(\hat{p}_1 - \hat{p}_2)$	**Confidence Interval** $\bar{x} \pm t^* \text{s.e.}(\bar{x})$ df = $n - 1$ **Paired Confidence Interval** $\bar{d} \pm t^* \text{s.e.}(\bar{d})$ df = $n - 1$

Two Population Means	
General	**Pooled**
Parameter $\mu_1 - \mu_2$	**Parameter** $\mu_1 - \mu_2$
Statistic $\bar{x}_1 - \bar{x}_2$	**Statistic** $\bar{x}_1 - \bar{x}_2$
Standard Error $\text{s.e.}(\bar{x}_1 - \bar{x}_2) = \sqrt{\frac{s_1^2}{n_1} + \frac{s_2^2}{n_2}}$	**Standard Error** pooled $\text{s.e.}(\bar{x}_1 - \bar{x}_2) = s_p\sqrt{\frac{1}{n_1} + \frac{1}{n_2}}$ where $s_p = \sqrt{\frac{(n_1-1)s_1^2 + (n_2-1)s_2^2}{n_1 + n_2 - 2}}$
Confidence Interval $(\bar{x}_1 - \bar{x}_2) \pm t^*(\text{s.e.}(\bar{x}_1 - \bar{x}_2))$ df = $\min(n_1 - 1, n_2 - 1)$	**Confidence Interval** $(\bar{x}_1 - \bar{x}_2) \pm t^*(\text{pooled s.e.}(\bar{x}_1 - \bar{x}_2))$ df = $n_1 + n_2 - 2$

Additional Notes

A place to ... jot down questions you may have and ask during office hours, take a few extra notes, write out an extra practice problem or summary completed in lecture, create your own short summary about this chapter.

Stat 250 Gunderson Lecture Notes
Learning about the Difference in Population Proportions

Part 3: Testing about a Difference in Population Proportions
Chapter 12: Section 3 HT Module 2

12.3 HT Module 2: Testing Hypotheses about the Difference in Two Population Proportions

- We have two populations from which independent samples are available, (or one population for which two groups can be formed using a categorical variable).
- The response variable is also **categorical** and we are interested in comparing the proportions for the two populations.

Let p_1 be the population proportion for the first population.
Let p_2 be the population proportion for the second population.

Parameter: the difference in the population proportions $p_1 - p_2$.

Sample estimate: the difference in the sample proportions $\hat{p}_1 - \hat{p}_2$.

Standard deviation of $\hat{p}_1 - \hat{p}_2$: $\text{s.d.}(\hat{p}_1 - \hat{p}_2) = \sqrt{\frac{p_1(1-p_1)}{n_1} + \frac{p_2(1-p_2)}{n_2}}$

Recall from Chapter 11 that the multiplier in the confidence interval was a z^* value. So we will be computing a Z test statistic for performing a significance test.

The standard error used in constructing the confidence interval for the difference between two population proportions is not the same as that used for the standardized z test statistic.

We will need to construct the null standard error, the standard error for the statistic when the null hypothesis is true. Let's start with what the hypotheses will look like.

Possible null and alternative hypotheses.

1. **H_0:** **versus H_a:**

2. **H_0:** **versus H_a:**

3. **H_0:** **versus H_a:**

Next we need to determine the test statistic and understand the conditions required for the test to be valid. The general form of the test statistic is:

$$\textbf{Test statistic} = \frac{\textbf{Sample statistic} - \textbf{Null value}}{\textbf{Standard error}}$$

In the case of two population proportions, if the null hypothesis is true, we have $p_1 - p_2 = 0$ or that the two population proportions are the same, $p_1 = p_2 = p$. What is a reasonable way to **estimate the common population proportion p**?

The general standard error for $\hat{p}_1 - \hat{p}_2$ is given by:

$$\text{s.e.}(\hat{p}_1 - \hat{p}_2) = \sqrt{\frac{\hat{p}_1(1-\hat{p}_1)}{n_1} + \frac{\hat{p}_2(1-\hat{p}_2)}{n_2}}$$

but if the null hypothesis is true, then $\hat{p}$ is the best estimate for each population proportion and should be used in the standard error.

So, the **null standard error** for $\hat{p}_1 - \hat{p}_2$ is given by:

And the corresponding test statistic is:

If the null hypothesis is true, this *z*-statistic will have a ______________ distribution. This distribution is used to find the *p*-value for the test.

Conditions: This test requires that the sample proportions are based on independent random samples from the two populations. Also, all of the quantities $n_1\hat{p}$, $n_1(1-\hat{p})$, $n_2\hat{p}$, and $n_2(1-\hat{p})$ be preferably at least 10. Note these are checked with the estimate of the common population proportion $\hat{p}$.

From the Stat 250 Formula Card we have the following summary:

Two Population Proportions	
Parameter	$p_1 - p_2$
Statistic	$\hat{p}_1 - \hat{p}_2$
Large-Sample *z*-Test	$z = \dfrac{\hat{p}_1 - \hat{p}_2}{\sqrt{\hat{p}(1-\hat{p})\left(\dfrac{1}{n_1} + \dfrac{1}{n_2}\right)}}$ where $\hat{p} = \dfrac{n_1\hat{p}_1 + n_2\hat{p}_2}{n_1 + n_2}$

Try It! Returning Lost Money $$

A random sample of college students was asked if they would return the money if they found a wallet on the street. They were also asked to report their gender so a comparison could be made between the two populations of male versus female college students.

Of the 93 women, 83 said they would, and of the 75 men, 53 said they would. Assume these students represent all college students.

Test the hypothesis that equal proportions of college men and women would say they would return the money versus the alternative hypothesis that a higher proportion of women would do so. Use a 5% significance level and show all steps.

Additional Notes

A place to ... jot down questions you may have and ask during office hours, take a few extra notes, write out an extra practice problem or summary completed in lecture, create your own short summary about this chapter.

Stat 250 Gunderson Lecture Notes
Learning about a Population Mean

Part 1: Distribution for a Sample Mean
Chapter 9: Section 6, SD Module 3 and Sections 9 and 10

Recall Parameters, Statistics and Sampling Distributions

We go back to the scenario where we have one population of interest but now the response being measured is **quantitative** (not categorical). We want to learn about the value of the **population mean** μ. We take a random sample and use the sample statistic, the **sample mean** $\overline{X}$, to estimate the parameter. When we do this, the sample mean may not be equal to the population mean, in fact, it could change every time we take a new random sample.

So recall that **a statistic is a random variable** and **it will have a probability distribution**. This probability distribution is called the **sampling distribution** of the statistic.

We turn to understanding the **sampling distribution of the sample mean** which will be used to construct a confidence interval estimate for the population mean and to test hypotheses about the value of a population mean.

9.5 SD Module 3: Sampling Distribution for One Sample Mean

Many responses of interest are measurements – height, weight, distance, reaction time, scores. We want to learn about a population mean and we will do so using the information provided from a sample from the population.

Example: How many hours per week do you work?

A poll was conducted by The Heldrich Center for Workforce Development (at Rutgers University). A probability sample of 1000 workers resulted in a mean number of hours worked per week of 43.

Population =

Parameter =

Sample =

Statistic =

Can anyone say how close this observed sample mean $\overline{x}$ of 43 is to the population mean μ?

If we were to take another random sample of the same size, would we get the same value for the sample mean?

So what are the possible values for the sample mean $\overline{x}$ if we took many random samples of the same size from this population? What would the distribution of the possible $\overline{x}$ values look like?

What can we say about the **distribution of the sample mean**?

Distribution of the Sample Mean – Main Results

Let μ = mean for the population of interest.
Let σ = standard deviation for the population of interest.
Let $\overline{x}$ = the sample mean for a random sample of size *n*.

If all possible random samples of the same size *n* are taken and $\overline{x}$ is computed for each, then ...

- The average of all of the possible sample mean values (from all possible random samples of the same size *n*)

  is equal to ______________________________.

  Thus the sample mean is an ___________________ estimator of the population mean.

- The standard deviation of all of the possible sample mean values is equal to the original population standard deviation divided by $\sqrt{n}$.

Standard deviation of the sample mean is given by: s.d.($\overline{x}$ **) =** $\frac{\sigma}{\sqrt{n}}$

What about the shape of the sampling distribution? The first two bullets above provide what the mean and the standard deviation are for the possible sample mean values. The final two bullets tell us that the shape of the distribution will be (approximately) normal.

- If the parent (original) **population has a normal distribution**, then the distribution of the possible values of $\overline{x}$, the sample mean, is **normal**.

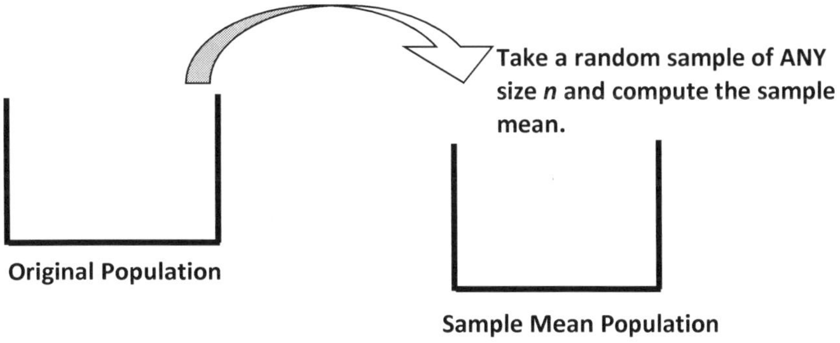

- If the parent (original) **population is not necessarily normally distributed** but the **sample size** n **is large**, then the distribution of the possible values of $\overline{x}$, the sample mean is ***approximately* normal.**

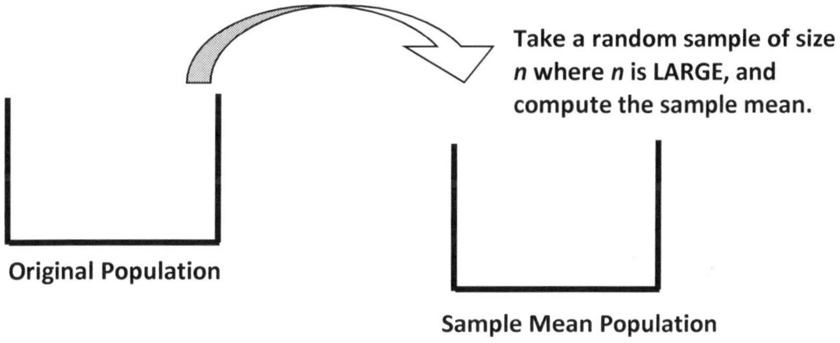

This last result is called the ________________________________ .

The CLT is formally presented in Section 9.10 of your text (starting on page 345 of the textbook). The **C** in CLT is for **CENTRAL**. The CLT is an important or central result in statistics. As it turns out, all normal curve approximations for the big five statistics are really applications of the CLT and can be derived using the results from Section 8.8.

The Stat 250 formula card summarizes distribution of a sample mean as follows:

Sample Means

Mean: $E(\overline{X}) = \mu_{\overline{X}} = \mu$

Standard Deviation: $s.d.(\overline{X}) = \sigma_{\overline{X}} = \frac{\sigma}{\sqrt{n}}$

Sampling Distribution of $\overline{X}$

If *X* has the $N(\mu, \sigma)$ distribution, then $\overline{X}$ is $N\left(\mu, \frac{\sigma}{\sqrt{n}}\right)$.

If *X* follows *any* distribution with mean μ and standard deviation σ and *n* is large, then $\overline{X}$ is *approximately* $N\left(\mu, \frac{\sigma}{\sqrt{n}}\right)$.

This last result is the **Central Limit Theorem**

Try It! ACT scores

Suppose that ACT scores for students follow a ***normal model*** with mean of 18.6 and standard deviation of 5.9.

a. What is the probability that a ***single student*** randomly selected will score 21 or higher?

b. Suppose two students will be randomly selected. What is the probability they will both score 21 or higher?

c. A random sample of 50 students is obtained. What is the probability that the ***mean score*** for these 50 students will be 21 or higher?

d. What if the normal distribution assumption were not given? Would your answer to part c change? Explain.

Try It! Actual Flight Times

Suppose the random variable *X* represents the actual flight time (in minutes) for Delta Airlines flights from Cincinnati to Tampa follows a *uniform distribution over the range of 110 minutes to 130 minutes.*

a. Sketch the distribution for *X* (include axes labels and some values on the axes).

b. Suppose we were to repeatedly take a random sample of size 100 from this distribution and compute the sample mean for each sample. What would the *histogram of the sample mean values* look like? Provide a smoothed out sketch of the distribution of the sample mean, include all details that you can.

Try It! True or False

Determine whether each of the following statements is True or False. A true statement is always true. Clearly circle your answer.

a. The central limit theorem is important in statistics because for a large random sample, it says the sampling distribution of the sample mean is approximately normal.
   True ***False***

b. The sampling distribution of a parameter is the distribution of the parameter value if repeated random samples are obtained.
   True ***False***

More on the Standard Deviation of $\overline{X}$

The standard deviation of the sample mean is given by: s.d.($\overline{x}$) = $\frac{\sigma}{\sqrt{n}}$

This quantity would give us an idea about how far apart a sample mean $\overline{x}$ and the true population mean μ are expected to be on average.

We can ***interpret*** **the standard deviation of the sample mean** as **approximately** the **average distance** of the possible sample mean value<u>s</u> (for repeated samples of the same size *n*) from the **true population mean** μ.

Note: If the sample size increases, the standard deviation decreases, which says the possible sample mean values will be closer to the true population mean (on average).

The **s.d.($\overline{x}$)** is a measure of the ***accuracy of the process*** of using a sample mean to estimate the population mean. This quantity $\frac{\sigma}{\sqrt{n}}$ does not tell us exactly how far away a particular observed $\overline{x}$ value is from μ.

In practice, the population standard deviation σ is rarely known, so the sample standard deviation *s* is used. As with proportions, when making this substitution we call the result the **standard error of the mean s.e.($\overline{x}$)** = $\frac{s}{\sqrt{n}}$. This terminology makes sense, because this is a measure of how much, on ***average***, the sample mean is in error as an estimate of the population mean.

Standard error of the sample mean is given by: s.e.($\overline{x}$) = $\frac{s}{\sqrt{n}}$

This quantity is an <u>*estimate*</u> of the standard deviation of $\overline{x}$.

So we can ***interpret*** **the standard error of the sample mean** as <u>**estimating**</u>, **approximately**, the **average distance** of the possible $\overline{x}$ value<u>s</u> (for repeated samples of the same size *n*) from the **true population mean** μ.

Moreover, we can use this standard error to create a range of values that we are very confident will contain the true population mean μ, namely, $\overline{x} \pm$(few)s.e.($\overline{x}$). This is the basis for confidence interval for the population mean μ, discussed more in the Chapter 11.

9.9 Preparing for Statistical Inference: Standardized Statistics

In our ACT Scores and Actual Flight Times examples earlier, we have already constructed and used a **standardized *z*-statistic for a sample mean.**

$z = \dfrac{\bar{x}-\mu}{\sigma/\sqrt{n}}$ has (approximately) a standard normal distribution $N(0,1)$.

Dilemma = ______________________________________

If we replace the population standard deviation σ with the sample standard deviation s, then $\dfrac{\bar{x}-\mu}{s/\sqrt{n}}$ won't be approximately $N(0,1)$; instead it has a

____

Student's *t*-Distribution

A little about the family of t-distributions ...

- They are symmetric, unimodal, centered at 0.
- They are flatter with heavier tails compared to the $N(0,1)$ distribution.
- As the degrees of freedom (df) increases ... the *t* distribution approaches the $N(0,1)$ distribution.
- We can still use the ideas about standard scores for a frame of reference.
- Tables A.2 and A.3 summarize percentiles for various t-distributions

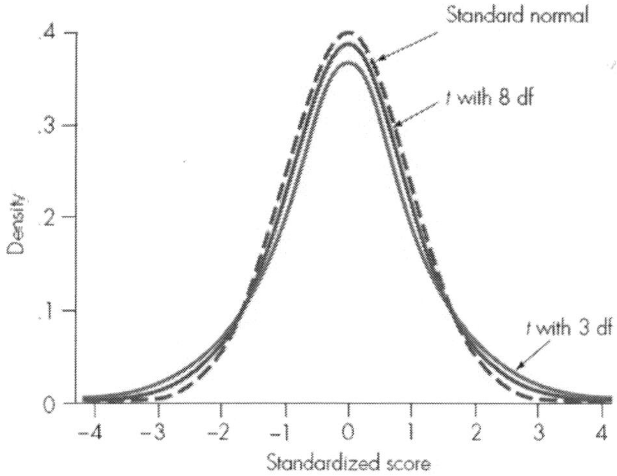

Figure 9.10 ■ *t*-distributions with df = 3, df = 8, and a standard normal distribution

From Utts, Jessica M. and Robert F. Heckard. Mind on Statistics, Fourth Edition. 2012. Used with permission.

We will see more on t-distributions in Chapters 11 and 13 when we do inference about population mean(s).

9.10 Generalizations beyond the Big Five

The **sampling distribution of a statistic** is the distribution of possible values of the statistic for repeated samples of the same size from a population.

So far we have discussed the sampling distribution of a sample proportion (SD Module 1), the sampling distribution of the difference between two sample proportions (SD Module 2), and the sampling distribution of the sample mean (SD Module 3). In all cases, under specified conditions the sampling distribution was *approximately* normal.

Every statistic has a sampling distribution, but the appropriate distribution may not always be normal, or even be approximately bell-shaped.

You can construct an approximate sampling distribution for any statistic by actually taking repeated samples of the same size from a population and constructing a histogram for the values of the statistic over the many samples.

The web applet examined in your computer labs could simulate the sampling distribution of the median, the range, and many other statistics. Example 9.15 on pages 347-348 of your text also shows a good example.

Additional Notes

A place to ... jot down questions you may have and ask during office hours, take a few extra notes, write out an extra practice problem or summary completed in lecture, create your own short summary about this chapter.

Stat 250 Gunderson Lecture Notes
Learning about a Population Mean

Part 2: Confidence Interval for a Population Mean
Chapter 11: Sections 1 and 2, CI Module 3

Do not put faith in what statistics say until you have carefully considered what they do not say. --*William W. Watt*

11.1 Introduction to Confidence Intervals for Means

Chapter 10 introduced us to **confidence intervals for estimating a population proportion and the difference between two population proportions**. Recall it is important to understand how to interpret an interval and how to interpret what the confidence level really means.

- The **interval provides a range of reasonable values** for the parameter with an associated high level of confidence. For example we can say, "We are 95% confident that the proportion of Americans who do not get enough sleep at night is somewhere between 0.325 to 0.395, based on a random sample of n = 935 American adults.
- The **95% confidence level describes our confidence in the procedure** we used to make the interval. If we repeated the procedure many times, we would expect about 95% of the intervals to contain the population parameter.

11.2 CI Module 3: Confidence Interval for a Population Mean μ

Consider a study on the **design of a highway sign**. A question of interest is: What is the **mean** maximum distance at which drivers are able to read the sign?

Data: A highway safety researcher will take a random sample of n = 16 drivers and measure the maximum distances (in feet) at which each can read the sign.

Population parameter:

μ = _________ **mean** maximum distance to read the sign for _________________

Sample estimate

$\bar{x}$ = _________ **mean** maximum distance to read the sign for _________________

But we know the sample estimate $\bar{x}$ may not equal μ, in fact, the possible $\bar{x}$ values vary from sample to sample. Because the sample mean is computed from a random sample, then it is a random variable, with a probability distribution.

Sampling Distribution of the sample mean

If $\bar{x}$ is the sample mean for a random sample of size *n* from a population with a normal model, then the distribution of the sample mean is:

Central Limit Theorem

If $\bar{x}$ is the sample mean for a random sample of size *n* from a population with any model, with mean, μ, and standard deviation σ, then when *n* is large, then the sampling distribution of the sample mean is ***approximately***:

So the possible $\bar{x}$ values vary normally around μ with a standard deviation of $\frac{\sigma}{\sqrt{n}}$. The standard deviation of the sample mean, $\frac{\sigma}{\sqrt{n}}$, is roughly the average distance of the possible sample mean values from the population mean μ. Since we don't know the population standard deviation σ, we will use the sample standard deviation s, resulting in the standard error of the sample mean.

Standard Error of the Sample mean

s.e.($\bar{x}$) = where *s* = sample standard deviation

The standard error of $\bar{x}$ estimates, roughly, the average distance of the possible $\bar{x}$ values from μ. The possible $\bar{x}$ values result from considering all possible random samples of the same size *n* from the same population.

So we have our estimate of the population mean, the sample mean $\bar{x}$, and we have its standard error. To make our confidence interval, we need to know the multiplier.

Sample Estimate $\pm$ Multiplier x Standard error

The **multiplier for a confidence interval for the population mean is denoted by *t****, which is the value in a **Student's t distribution with df = *n* – 1** such that the area between –*t* and *t* equals the desired confidence level. The value of *t** will be found using Table A.2. First let's give the formal result.

One-sample t Confidence Interval for μ

$$\bar{x} \pm t^{*} \text{s.e.}(\bar{x})$$

where t^* is an appropriate value for a $t(n-1)$ distribution.

This interval requires we have a random sample from a normal population. If the sample size is large ($n > 30$), the assumption of normality is not so crucial and the result is approximate.

Important items:

- be sure to check the conditions
- know how to interpret the confidence interval
- be able to explain what the confidence level of say 95% really means

Try It! Using Table A.2 to find *t**

(a) Find t^* for a 90% confidence interval based on $n = 12$ observations.

(b) Find t^* for a 95% confidence interval based on $n = 30$ observations.

(c) Find t^* for a 95% confidence interval based on $n = 54$ observations.

(d) What happens to the value of t^* as the sample size (and thus the degrees of freedom) gets larger?

Table A.2 t^* Multipliers for Confidence Intervals and Rejection Region Critical Values

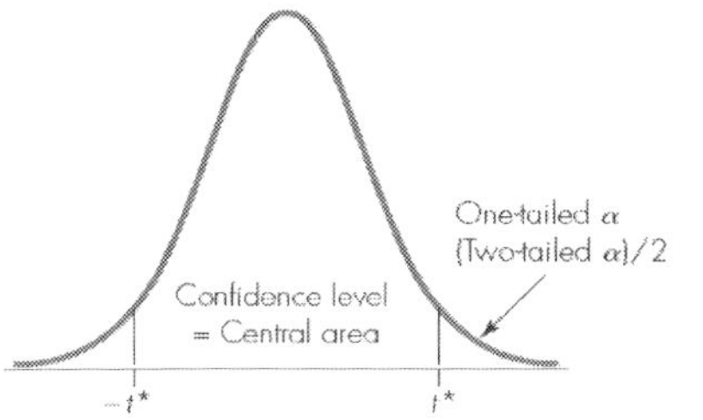

	Confidence Level						
df	*.80*	*.90*	*.95*	*.98*	*.99*	*.998*	*.999*
1	3.08	6.31	12.71	31.82	63.66	318.31	636.62
2	1.89	2.92	4.30	6.96	9.92	22.33	31.60
3	1.64	2.35	3.18	4.54	5.84	10.21	12.92
4	1.53	2.13	2.78	3.75	4.60	7.17	8.61
5	1.48	2.02	2.57	3.36	4.03	5.89	6.87
6	1.44	1.94	2.45	3.14	3.71	5.21	5.96
7	1.41	1.89	2.36	3.00	3.50	4.79	5.41
8	1.40	1.86	2.31	2.90	3.36	4.50	5.04
9	1.38	1.83	2.26	2.82	3.25	4.30	4.78
10	1.37	1.81	2.23	2.76	3.17	4.14	4.59
11	1.36	1.80	2.20	2.72	3.11	4.02	4.44
12	1.36	1.78	2.18	2.68	3.05	3.93	4.32
13	1.35	1.77	2.16	2.65	3.01	3.85	4.22
14	1.35	1.76	2.14	2.62	2.98	3.79	4.14
15	1.34	1.75	2.13	2.60	2.95	3.73	4.07
16	1.34	1.75	2.12	2.58	2.92	3.69	4.01
17	1.33	1.74	2.11	2.57	2.90	3.65	3.97
18	1.33	1.73	2.10	2.55	2.88	3.61	3.92
19	1.33	1.73	2.09	2.54	2.86	3.58	3.88
20	1.33	1.72	2.09	2.53	2.85	3.55	3.85
21	1.32	1.72	2.08	2.52	2.83	3.53	3.82
22	1.32	1.72	2.07	2.51	2.82	3.50	3.79
23	1.32	1.71	2.07	2.50	2.81	3.48	3.77
24	1.32	1.71	2.06	2.49	2.80	3.47	3.75
25	1.32	1.71	2.06	2.49	2.79	3.45	3.73
26	1.31	1.71	2.06	2.48	2.78	3.43	3.71
27	1.31	1.70	2.05	2.47	2.77	3.42	3.69
28	1.31	1.70	2.05	2.47	2.76	3.41	3.67
29	1.31	1.70	2.05	2.46	2.76	3.40	3.66
30	1.31	1.70	2.04	2.46	2.75	3.39	3.65
40	1.30	1.68	2.02	2.42	2.70	3.31	3.55
50	1.30	1.68	2.01	2.40	2.68	3.26	3.50
60	1.30	1.67	2.00	2.39	2.66	3.23	3.46
70	1.29	1.67	1.99	2.38	2.65	3.21	3.44
80	1.29	1.66	1.99	2.37	2.64	3.20	3.42
90	1.29	1.66	1.99	2.37	2.63	3.18	3.40
100	1.29	1.66	1.98	2.36	2.63	3.17	3.39
1000	1.282	1.646	1.962	2.330	2.581	3.098	3.300
Infinite	1.281	1.645	1.960	2.326	2.576	3.090	3.291
Two-tailed α	.20	.10	.05	.02	.01	.002	.001
One-tailed α	.10	.05	.025	.01	.005	.001	.0005

Note that the t-distribution with infinite df is the standard normal distribution.

From Utts, Jessica M. and Robert F. Heckard. Mind on Statistics, Fourth Edition. 2012. Used with permission.

Try It! Confidence Interval for the Mean Maximum Distance

Recall the study on the design of a highway sign. The researcher wanted to learn about the **mean** maximum distance at which drivers are able to read the sign. The researcher took a *random sample* of n = 16 drivers and measured the maximum distances (in feet) at which each can read the sign. The data are provided below.

440	**490**	**600**	**540**	**540**	**600**	**240**	**440**
360	**600**	**490**	**400**	**490**	**540**	**440**	**490**

a. Verify the necessary conditions for computing a confidence interval for the population mean distance. We are told that the sample was a random sample so we just need to check if a normal model for the response 'max distance' for the population is reasonable.

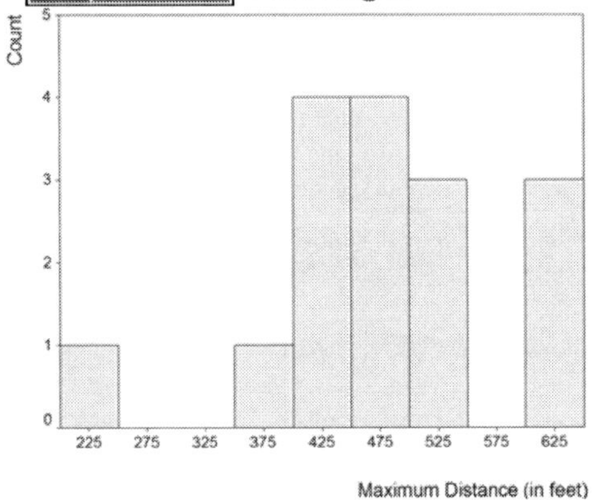

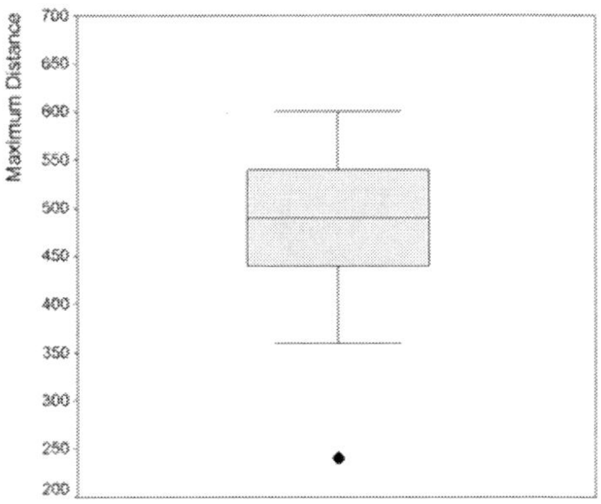

Comments:

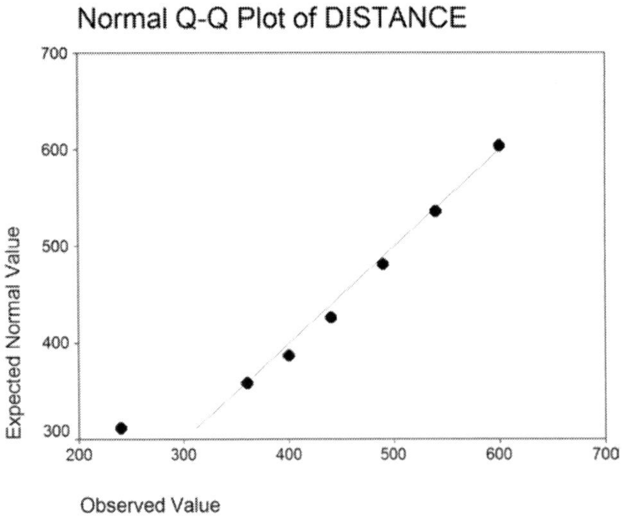

440	**490**	**600**	**540**	**540**	**600**	[illegible]	**440**
360	**600**	**490**	**400**	**490**	**540**	**440**	**490**

b. Compute the sample mean maximum distance and the standard error (without the outlier).

c. Use a 95% confidence interval to estimate the population mean maximum distance at which all drivers can read the sign. Write a paragraph that interprets this interval and the confidence level.

PD-INEL

One-Sample Statistics

	N	Mean	Std. Deviation	Std. Error Mean
DISTANCE	15	497.3333	73.43283	18.96028

One-Sample Test

	Test Value = 0					
					95% Confidence Interval of the Difference	
	t	df	Sig. (2-tailed)	Mean Difference	Lower	Upper
DISTANCE	26.230	14	.000	497.3333	456.6676	537.9991

Here is the summary of the Confidence intervals for the big 5 parameters covered in Chapters 10 and 11. We have now covered the one and two population proportion scenarios and the one population mean scenario

Population Proportion	Two Population Proportions	Population Mean
Parameter p	**Parameter** $p_1 - p_2$	**Parameter** μ
Statistic $\hat{p}$	**Statistic** $\hat{p}_1 - \hat{p}_2$	**Statistic** $\bar{x}$
Standard Error $\text{s.e.}(\hat{p}) = \sqrt{\frac{\hat{p}(1-\hat{p})}{n}}$	**Standard Error** $\text{s.e.}(\hat{p}_1 - \hat{p}_2) = \sqrt{\frac{\hat{p}_1(1-\hat{p}_1)}{n_1} + \frac{\hat{p}_2(1-\hat{p}_2)}{n_2}}$	**Standard Error** $\text{s.e.}(\bar{x}) = \frac{s}{\sqrt{n}}$
Confidence Interval $\hat{p} \pm z^* \text{s.e.}(\hat{p})$ **Conservative Conf. Interval** $\hat{p} \pm \frac{z^*}{2\sqrt{n}}$ **Sample Size** $n = \left(\frac{z^*}{2m}\right)^2$	**Confidence Interval** $(\hat{p}_1 - \hat{p}_2) \pm z^* \text{s.e.}(\hat{p}_1 - \hat{p}_2)$	**Confidence Interval** $\bar{x} \pm t^* \text{s.e.}(\bar{x})$ df = $n-1$ **Paired Confidence Interval** $\bar{d} \pm t^* \text{s.e.}(\bar{d})$ df = $n-1$

Two Population Means

General		Pooled	
Parameter	$\mu_1 - \mu_2$	**Parameter**	$\mu_1 - \mu_2$
Statistic	$\bar{x}_1 - \bar{x}_2$	**Statistic**	$\bar{x}_1 - \bar{x}_2$
Standard Error $\text{s.e.}(\bar{x}_1 - \bar{x}_2) = \sqrt{\frac{s_1^2}{n_1} + \frac{s_2^2}{n_2}}$		**Standard Error** pooled $\text{s.e.}(\bar{x}_1 - \bar{x}_2) = s_p\sqrt{\frac{1}{n_1} + \frac{1}{n_2}}$ where $s_p = \sqrt{\frac{(n_1-1)s_1^2 + (n_2-1)s_2^2}{n_1+n_2-2}}$	
Confidence Interval $(\bar{x}_1 - \bar{x}_2) \pm t^*(\text{s.e.}(\bar{x}_1 - \bar{x}_2))$ df = $\min(n_1 - 1, n_2 - 1)$		**Confidence Interval** $(\bar{x}_1 - \bar{x}_2) \pm t^*(\text{pooled s.e.}(\bar{x}_1 - \bar{x}_2))$ df = $n_1 + n_2 - 2$	

Additional Notes

A place to ... jot down questions you may have and ask during office hours, take a few extra notes, write out an extra practice problem or summary completed in lecture, create your own short summary about this chapter.

Stat 250 Gunderson Lecture Notes
Learning about a Population Mean

Part 3: Testing about a Population Mean
Chapter 13: Sections 1, 2, and 5 HT Module 3

13.1 Introduction to Hypothesis Tests for Means

Chapter 12 introduced us to the **logic and steps of hypothesis testing for learning about a population proportion and for the difference between two population proportions**. Recall the big idea that we declare "statistical significance" and reject the null hypothesis if the *p*-value is less than or equal to the significance level α. In Chapter 13 we extend these ideas to testing about means, focusing first on hypothesis testing about a single **population mean**.

Page 499 of your textbook provides **three notes of caution** that are worthy of reviewing here:

1. **Inference is only valid if the sample is representative** of the population for the question of interest.
2. **Hypotheses and conclusions apply to the larger population(s)** represented by the sample(s).
3. **If the distribution of a quantitative variable is highly skewed**, we should consider analyzing the median rather than the mean. Methods for testing hypotheses about medians are a special case of **nonparametric methods**, two examples of which will be covered briefly in Section 16.3.

Next let's review the **Basic Steps in Any Hypothesis Test** from Chapter 12.

Step 1: Determine the null and alternative hypotheses.
The hypotheses are statements about the population(s), not the sample(s).
The null hypothesis defines a specific value of a population parameter, called the **null value**.

Step 2: Verify necessary data conditions, and if met, summarize the data into an appropriate test statistic.

A relevant statistic is calculated from sample information and summarized into a "test statistic." We measure the difference between the sample statistic and the null value using the standardized statistic:

$$\frac{\textbf{Sample statistic} - \textbf{Null value}}{\textbf{(Null) standard error}}$$

For hypotheses about **proportions**, the standardized statistic is called a ______________ and the _________________________ is used to find the *p*-value.

For hypotheses about **means**, the standardized statistic is called a ______________ and the _________________________ is used to find the *p*-value.

Step 3: Assuming the null hypothesis is true, find the *p*-value.

A *p*-value is computed based on the standardized "test statistic." The *p*-value is calculated by temporarily assuming the null hypothesis to be true and then calculating the probability that the test statistic could be as large in magnitude as it is (or larger) in the direction(s) specified by the alternative hypothesis.

Step 4: Decide if the result is statistically significant based on the *p*-value.

Based on the *p*-value, we either reject or fail to reject the null hypothesis. The most commonly used criterion (level of significance) is that we reject the null hypothesis when the *p*-value is less than or equal to the significance level (generally 0.05). In many research articles, *p*-values are simply reported and readers are left to draw their own conclusions. Remember that a *p*-value measures the strength of the evidence against the null hypothesis and the smaller the *p*-value, the stronger the evidence against the null (and for the alternative).

The *Beauty* of *p*-values: Suppose the significance level α is set at 5% for testing H_0: status quo versus H_a: the "new theory".

If *p*-value is...	Statistical Decision	Feasible Conclusion about the "New Theory"
0.462	Fail to Reject H_0	
0.063	Fail to Reject H_0	
0.041	Reject H_0	
0.003	Reject H_0	

Step 5: Report the conclusion in the context of the situation.

The decision is to reject or fail to reject the null hypothesis, but the conclusion should go back to the original question of interest being asked. It should be stated in terms of the particular scenario or situation.

13.2 HT Module 3: Testing Hypotheses about One Population Mean μ

Situation:
We have **one population** and a **response that is quantitative**.
We wish to test about the value of the **mean response for the population** μ.
The data are assumed to be a **random sample**.
The response is assumed to be **normally distributed** for the population (but if the sample size is large, this condition is less crucial).

Step 1: Determine the null and alternative hypotheses.

1. H_0: versus H_a:

2. H_0: versus H_a:

3. H_0: versus H_a:

Step 2: Verify necessary data conditions, and if met, summarize the data into an appropriate test statistic.

How would you check the conditions as stated in the scenario above?

$$\textbf{Test statistic} = \frac{\textbf{Sample statistic} - \textbf{Null value}}{\textbf{Standard error}}$$

If H_0 is true, this test statistic has a ______________ distribution.

Step 3: Assuming the null hypothesis is true, find the *p*-value.

Table 13.1 Alternative Hypothesis Regions and *p*-Value Areas

Statement of H_a		*p*-Value Area	*t*-Curve Region
$\mu < \mu_0$	(less than)	Area to the left of t (even if $t > 0$)	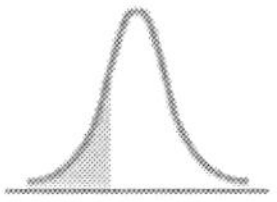
$\mu > \mu_0$	(greater than)	Area to the right of t (even if $t < 0$)	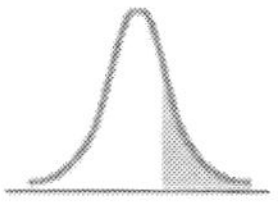
$\mu \neq \mu_0$	(not equal)	$2 \times$ area to the right of $\lvert t \rvert$	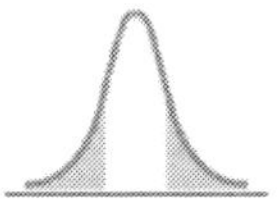

From Utts, Jessica M. and Robert F. Heckard. Mind on Statistics, Fourth Edition. 2012. Used with permission.

Steps for finding a *p*-value ...

- **Draw the distribution for the test statistic under H_0**
  For t tests it will be a t-distribution with a certain df.
- **Locate the observed test statistic value on the axis.**
- **Shade in the area that corresponds to the *p*-value.**
  Look at the alternative hypothesis for the direction of extreme.
- **Use the appropriate table to find (bounds for) the *p*-value.**
  For t tests we use Table A.3.

Step 4: Decide whether or not the result is statistically significant based on the *p*-value.

The level of significance α is selected in advance. **We reject the null hypothesis if the *p*-value is less than or equal to α.**. In this case, we say the results are statistically significant at the level α.

Step 5: Report the conclusion in the context of the situation.

Once the decision is made, a conclusion in the context of the problem can be stated.

From the Stat 250 formula card:

Population Mean	
Parameter	μ
Statistic	$\bar{x}$
Standard Error $\text{s.e.}(\bar{x}) = \dfrac{s}{\sqrt{n}}$	
One-Sample *t*-Test $t = \dfrac{\bar{x} - \mu_0}{\text{s.e.}(\bar{x})} = \dfrac{\bar{x} - \mu_0}{s/\sqrt{n}}$	df = $n - 1$

TABLE A.3 One-Sided p-Values for Significance Tests Based on a t-Statistic

- A p-value in the table is the area to the right of t.
- Double the value if the alternative hypothesis is two-sided (not equal).

	Absolute Value of t-Statistic							
df	*1.28*	*1.50*	*1.65*	*1.80*	*2.00*	*2.33*	*2.58*	*3.00*
1	0.211	0.187	0.173	0.161	0.148	0.129	0.118	0.102
2	0.164	0.136	0.120	0.107	0.092	0.073	0.062	0.048
3	0.145	0.115	0.099	0.085	0.070	0.051	0.041	0.029
4	0.135	0.104	0.087	0.073	0.058	0.040	0.031	0.020
5	0.128	0.097	0.080	0.066	0.051	0.034	0.025	0.015
6	0.124	0.092	0.075	0.061	0.046	0.029	0.021	0.012
7	0.121	0.089	0.071	0.057	0.043	0.026	0.018	0.010
8	0.118	0.086	0.069	0.055	0.040	0.024	0.016	0.009
9	0.116	0.084	0.067	0.053	0.038	0.022	0.015	0.007
10	0.115	0.082	0.065	0.051	0.037	0.021	0.014	0.007
11	0.113	0.081	0.064	0.050	0.035	0.020	0.013	0.006
12	0.112	0.080	0.062	0.049	0.034	0.019	0.012	0.006
13	0.111	0.079	0.061	0.048	0.033	0.018	0.011	0.005
14	0.111	0.078	0.061	0.047	0.033	0.018	0.011	0.005
15	0.110	0.077	0.060	0.046	0.032	0.017	0.010	0.004
16	0.109	0.077	0.059	0.045	0.031	0.017	0.010	0.004
17	0.109	0.076	0.059	0.045	0.031	0.016	0.010	0.004
18	0.108	0.075	0.058	0.044	0.030	0.016	0.009	0.004
19	0.108	0.075	0.058	0.044	0.030	0.015	0.009	0.004
20	0.108	0.075	0.057	0.043	0.030	0.015	0.009	0.004
21	0.107	0.074	0.057	0.043	0.029	0.015	0.009	0.003
22	0.107	0.074	0.057	0.043	0.029	0.015	0.009	0.003
23	0.107	0.074	0.056	0.042	0.029	0.014	0.008	0.003
24	0.106	0.073	0.056	0.042	0.028	0.014	0.008	0.003
25	0.106	0.073	0.056	0.042	0.028	0.014	0.008	0.003
26	0.106	0.073	0.055	0.042	0.028	0.014	0.008	0.003
27	0.106	0.073	0.055	0.042	0.028	0.014	0.008	0.003
28	0.106	0.072	0.055	0.041	0.028	0.014	0.008	0.003
29	0.105	0.072	0.055	0.041	0.027	0.013	0.008	0.003
30	0.105	0.072	0.055	0.041	0.027	0.013	0.008	0.003
40	0.104	0.071	0.053	0.040	0.026	0.012	0.007	0.002
50	0.103	0.070	0.053	0.039	0.025	0.012	0.006	0.002
60	0.103	0.069	0.052	0.038	0.025	0.012	0.006	0.002
70	0.102	0.069	0.052	0.038	0.025	0.011	0.006	0.002
80	0.102	0.069	0.051	0.038	0.024	0.011	0.006	0.002
90	0.102	0.069	0.051	0.038	0.024	0.011	0.006	0.002
100	0.102	0.068	0.051	0.037	0.024	0.011	0.006	0.002
1000	0.100	0.067	0.050	0.036	0.023	0.010	0.005	0.001
Infinite	0.1003	0.0668	0.0495	0.0359	0.0228	0.0099	0.0049	0.0013

Note that the t-distribution with infinite df is the standard normal distribution.

Try It! Using Table A.3 to find a *p*-value for a one-sided test

We are testing H_0: $\mu = 0$ versus H_a: $\mu > 0$ with $n = 15$ observations and the observed test statistic is $t = 1.97$

- **Draw the distribution for the test statistic under H_0**

- **Locate the observed test statistic value on the axis.**
- **Shade in the area that corresponds to the *p*-value.**
  Look at the alternative hypothesis for the direction of extreme.
- **Use the appropriate table to find (bounds for) the *p*-value.**
  For t tests we will use Table A.3.

Is the value of $t = 1.97$ significant at the 5% level? __________

At the 1% level? _________

Try It! Using Table A.3 to find a *p*-value for a two-sided test

We are testing H_0: $\mu = 64$ versus H_a: $\mu \neq 64$ with $n = 30$ observations and the observed test statistic is $t = 1.12$. How would you report the *p*-value for this test?

Try It! Classical Music

A researcher wants to test if HS students complete a maze more quickly while listening to classical music. For the general HS population, the time to complete the maze is assumed to follow a normal distribution with a mean of 40 seconds. Use a 5% significance level.

Define the parameter of interest: Let μ represent...

State the hypotheses: **H_0:**

H_a:

A random sample of 100 HS students are timed while listening to classical music. The mean time was 39.1 seconds and the standard deviation was 4 seconds. Conduct the test.

Are the results statistically significant at the 5% level? ____________________

State the conclusion at the 5% level in terms of the problem.

Comment about the assumptions required for this test to be valid:

Try It! Calcium Intake

A bone health study looked at the daily intake of calcium (mg) for 38 women. They are concerned that the mean calcium intake for the population of such women is not meeting the RDA level of 1200 mg, that is, the population mean is less than the 1200 mg level. They wish to test this theory using a 5% significance level.

a. State the hypotheses about the mean calcium intake for the population of such women.

H_0: ______________________ versus H_a: ______________________

One-Sample Statistics

	N	Mean	Std. Deviation	Std. Error Mean
CALTAKE	38	926.03	427.23	69.31

One-Sample Test

	Test Value = 1200					
					95% Confidence Interval of the Difference	
	t	df	Sig. (2-tailed)	Mean Difference	Lower	Upper
CALCIUM	-3.953	37	.00033	-273.97	-414.40	-133.55

b. Interpret the Std. Error of the mean (SEM):

c. Give the observed test statistic value: _____ = _______________
Interpret the this value in terms of a difference from the hypothesized mean of 1200.

d. Sketch a picture of the *p*-value in terms of an area under a distribution.

e. Give the *p*-value and the conclusion using a 5% significance level.

13.5 The Relationship between Significance Tests and Confidence Intervals

Earlier we discussed the using of confidence intervals to guide decisions. The main ideas:

- A confidence interval provides a **range of plausible (reasonable) values** for the parameter.
- The null hypothesis gives a **null value** for the parameter.
- **If this null value is one of the "reasonable" values** found in the confidence interval, the **null hypothesis would not be rejected**.
- **If this null value was not found in the confidence interval** of acceptable values for the parameter, then **the null hypothesis would be rejected**.

The summary given on page 521 is provided below.

> **Confidence Intervals and Tests with Two-Sided Alternatives**
>
> In testing one population mean or the difference in two population means with the hypotheses
>
> H_0: parameter = null value *and* H_a: parameter $\neq$ null value
>
> - If the null value is covered by a $(1 - \alpha)100\%$ confidence interval, the null hypothesis is not rejected, and the test is not statistically significant at level α.
> - If the null value is not covered by a $(1 - \alpha)100\%$ confidence interval, the null hypothesis is rejected, and the test is statistically significant at level α.
>
> Don't be confused by the cumbersome notation $(1 - \alpha)100\%$. The correspondence is between a 95% confidence interval and a significance level of $\alpha = .05$, between a 99% confidence interval and a significance level of $\alpha = .01$, and so on.

From Utts, Jessica M. and Robert F. Heckard. Mind on Statistics, Fourth Edition. 2012. Used with permission.

Notes:

(1) The alternative hypothesis should be **two-sided**. However, sometimes you can reason through the decision for a one-sided test (read page 521-522).

(2) The **significance level of the test should coincide with the confidence level** (e.g. α = 0.05 with a 95% confidence level). However, sometimes you can still determine the decision if these do not exactly correspond (see part (c) of the next Try It!).

(3) This relationship holds exactly for tests about a population mean or difference between two population means. In most cases, the correspondence will hold for tests about a population proportion or difference between two population proportions (read page 522).

Try It! Time Spent Watching TV

A study looked at the amount of time that teenagers are spending watching TV. Based on a representative sample, the 95% confidence interval for mean amount of time (in hours) spent watching TV on a weekend day was given as: 2.6 hours ± 2.1 hours. So the interval goes from 0.5 hours to 4.7 hours.

a. Test H_0: $\mu = 5$ hours versus H_a: $\mu \neq 5$ hours at $\alpha = 0.05$.

Reject H_0 Fail to reject H_0 Can't tell

Why?

b. Test H_0: $\mu = 4$ hours versus H_a: $\mu \neq 4$ hours at $\alpha = 0.05$.

Reject H_0 Fail to reject H_0 Can't tell

Why?

c. Test H_0: $\mu = 4$ hours versus H_a: $\mu \neq 4$ hours at $\alpha = 0.01$

Reject H_0 Fail to reject H_0 Can't tell

Why?

d. Test H_0: $\mu = 4$ hours versus H_a: $\mu \neq 4$ hours at $\alpha = 0.10$

Reject H_0 Fail to reject H_0 Can't tell

Why?

Try It! MBA grads Salaries

"It's a good year for MBA grads" was the title of an article. One of the parameters of interest was the population mean expected salary, μ (in dollars). A random sample of 1000 students who finished their MBA this year (from 129 business schools) resulted in a 95% confidence interval for μ of (83700, 84800).

a. What is the value of the sample mean? Include your units.

b. For each statement determine if it is true or false. ***Clearly circle your answer.***

If repeated samples of 1000 such students were obtained, we would expect 95% of the resulting intervals to contain the population mean.

True **False**

There is a 95% probability that the population mean lies between \$83,700 and \$84,800.

True **False**

c. The expected average earnings for such graduates in past year was \$76,100. Suppose we wish to test the following hypotheses at the *10% significance level*:

$$H_0: \mu = 76100 \text{ versus } H_a: \mu \neq 76100.$$

Our decision would be: **Fail to reject** H_0 **Reject** H_0 **can't tell**

Because ...

d. Several plots of the expected salary data were constructed to help verify some of the data conditions. A qq-plot is provided for checking the assumption that the response is normally distributed. This plot shows some departure from a straight line with a positive slope. Is this cause for concern that inference based on our confidence interval and hypothesis test would not be valid? Explain.

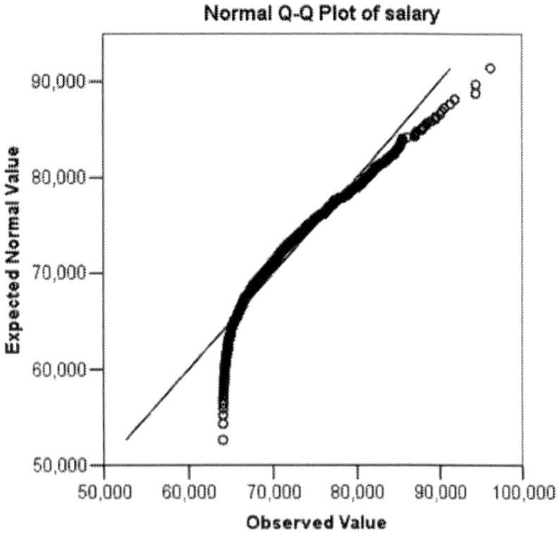

Additional Notes

A place to ... jot down questions you may have and ask during office hours, take a few extra notes, write out an extra practice problem or summary completed in lecture, create your own short summary about this chapter.

Stat 250 Gunderson Lecture Notes
Learning about a Population Mean Difference

Part 1: Distribution for a Sample Mean of Paired Differences
Chapter 9: Section 7, SD Module 4

The Paired Data Scenario

An important special case of a single mean of a population occurs when two quantitative variables are **collected in pairs**, and we desire information about the **difference** between the two variables.

Here are some ways that paired data can occur:

- Each person or unit is measured twice. The two measurements of the same characteristic or trait are made under different conditions. An example is measuring a quantitative response both before and after treatment.

- Similar individuals or units are paired prior to an experiment. During the experiment, each member of a pair receives a different treatment. The same quantitative response variable is measured for all individuals.

For paired data designs, it is the ***differences*** that we are interested in examining. By focusing on the differences we again have just one sample of observations (the differences). Sometimes you may see a "*d*" in the subscript of the mean to represent the **mean of the population of differences**: μ_d; and the data may be represented generically as: $d_1, d_2, \ldots, d_n$.

Here is an example of a paired data scenario from page 318 of your text.

Situation 4. Estimating the Mean of Paired Differences for a Quantitative Variable

- *Example research questions:*

  What is the mean difference in weights for freshmen at the beginning and end of the first semester?

  What is the mean difference in age between husbands and wives in Britain?

- *Population parameter:* μ_d = population mean of the differences in the two measurements.

- *Sample estimate:* $\overline{d}$ = the mean of the differences for a sample of the two measurements.

From Utts, Jessica M. and Robert F. Heckard. Mind on Statistics, Fourth Edition. 2012. Used with permission.

You might review pages 335-336 for the short discussion and example regarding a population mean for paired difference.

9.7 SD Module 4: Sampling Distribution for the Sample Mean of Paired Differences

The sampling distribution results for the mean of paired differences is really the same as that for a regular sample mean. Since the measurements are differences, the sample mean of the data, $\bar{x}$, is just written as $\bar{d}$, to emphasize that this is a paired design.

Freshmen Weight (Example 9.9 on page 337 of the textbook)
A study was conducted to learn about the average weight gain in the first year of college for students. A sample of 60 students resulted in an average weight gain of 4.2 pounds (over the first 12 weeks of college).

Population =

Parameter =

Sample =

Statistic =

Can anyone say how close this observed sample mean difference $\bar{d}$ of 4.2 pounds is to the true population mean difference μ_d? __________

If we were to take another random sample of the same size, would we get the same value for the sample mean difference? ________________________

So what are the possible values for the sample mean difference $\bar{d}$ if we took many random samples of the same size from this population? What would the distribution of the possible $\bar{d}$ values look like? What can we say about the **distribution of the sample mean difference**?

Distribution of the Sample Mean Difference – Main Results

Let μ_d = mean of the difference in the population of interest.
Let σ_d = standard deviation for the differences in the population of interest.
Let $\bar{d}$ = the sample mean of the differences for a random sample of size *n*.

- If the population of differences is normal (bell-shaped), and a random sample of any size is obtained, then the distribution of the sample mean difference $\bar{d}$ is also normal, with a mean of μ_d and a standard deviation of $s.d.(\bar{d}) = \frac{\sigma_d}{\sqrt{n}}$.
- If the population of differences is not normal (bell-shaped), but a *large* random sample of size *n* is obtained, then the distribution of the sample mean difference $\bar{d}$ is *approximately* normal, with a mean of μ_d and a standard deviation of $s.d.(\bar{d}) = \frac{\sigma_d}{\sqrt{n}}$.

Notes:

(1) An arbitrary level for what is 'large' enough has been 30. However, if any of the differences are extreme outliers, it is better to have a larger sample size.

(2) The standard deviation of $\bar{d}$ is a measure of the ***accuracy of the process*** of using a sample mean difference to estimate the population mean difference.

$$s.d.(\bar{d}) = \frac{\sigma_d}{\sqrt{n}}$$

(3) In practice, the population standard deviation σ_d is rarely known, so the sample standard deviation s_d is used instead. When making this substitution we call the result a **standard error**.

Standard error of the sample mean difference is given by:

$$\text{s.e.}(\bar{d}) = \frac{s_d}{\sqrt{n}}$$

We can ***interpret*** **the standard error of the sample mean difference** as **estimating**, **approximately**, the **average distance** of the possible $\bar{d}$ values (for repeated samples of the same size *n*) from the **population mean difference** μ_d.

Moreover, we can use this standard error of the sample mean difference to produce a range of values that we are very confident will contain the population mean difference μ_d, namely, $\bar{d} \pm$ (a few)s.e.($\bar{d}$). This is the basis for confidence interval for the population mean difference μ_d, discussed in Chapter 11.

Looking ahead:

Do you think the 'few' in the above expression will be a z^* value or a t^* value?

What degrees of freedom will be used?

We will use the standard error of the sample mean difference to compute a standardized test statistic for testing hypotheses about the population mean difference μ_d, namely,

$$\frac{\textbf{Sample statistic – Null value}}{\textbf{(Null) standard error}}.$$

This is the basis for testing about a population mean difference covered in Chapter 13.

Looking ahead:

Do you think the standardized test statistic will be a *z* statistic or a *t* statistic?

What do you think will be the most common null value?

H_0: μ_d = _______

Additional Notes

A place to ... jot down questions you may have and ask during office hours, take a few extra notes, write out an extra practice problem or summary completed in lecture, create your own short summary about this chapter.

Stat 250 Gunderson Lecture Notes
Learning about a Population Mean Difference

Part 2: Confidence Interval for a Population Mean of Paired Differences Chapter 11: Section 3, CI Module 4

11.3 CI Module 4: Confidence Interval for the Population Mean of Paired Differences μ_d

Recall that an important special case of a single mean of a population occurs when two quantitative variables are **collected in pairs**, and we desire information about the **difference** between the two variables. For paired data designs, it is the ***differences*** that we are interested in analyzing. By focusing on the differences we again have just one sample of observations (the differences) and are able to use the confidence interval for the population mean difference.

Notation:
Population Parameter: μ_d = population mean difference
Data: $d_1, d_2, ..., d_n$
Sample Estimate: $\bar{d}$ = sample mean difference
Standard Error: s.e.($\bar{d}$) = $\frac{s_d}{\sqrt{n}}$ (s_d = standard deviation of the sampled differences)

We use the sample estimate and its standard error to **form a confidence interval estimate** for the parameter using the following form:

Sample Estimate $\pm$ Multiplier x Standard error

The multiplier used will depend on the confidence level, the sample size, and the type of parameter being estimated. In this case, since we are estimating a single population mean, the multiplier will be a t^* value. Here is the summary for a paired data confidence interval:

One-sample *t* Confidence Interval for the Population Mean Difference μ_d

$$\bar{d} \pm t^* \text{s.e.}(\bar{d})$$

where t^* is the appropriate value for a $t(n-1)$ distribution.
Note that n is the number of pairs, or the number of differences.
This interval requires that the differences can be considered a random sample from a normal population. If the sample size is large, the assumption of normality is not so crucial and the result is approximate.

Try It! Changes in Reasoning Scores

Do piano lessons improve spatial-temporal reasoning of preschool children?
Data: The change in reasoning score, after piano lessons - before piano lessons, with larger values indicating better reasoning, for a random sample of n = 34 preschool children.

2	5	7	-2	2	7	4	1	0	7	3	4
3	4	9	4	5	2	9	6	0	3	6	-1
3	4	6	7	-2	7	-3	3	4	4		

(a) Display the data, summarize the distribution.
These data were entered into SPSS to produce the following histogram.

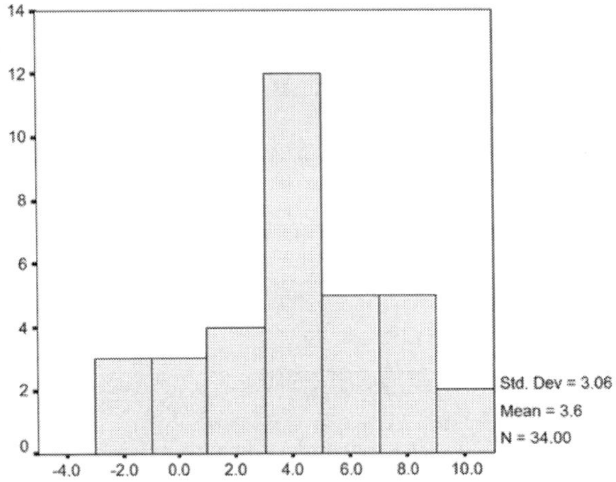

Change in Reasoning

Notes:
1. Diff = after – before so ...

2. Sample mean difference = ______

3. Normality of the response (the difference) for the population?

Some summary measures:

One-Sample Statistics

	N	Mean	Std. Deviation	Std. Error Mean
REASCORE	34	3.62	3.06	.52

(b) Give a 95% confidence interval for the population mean improvement in reasoning scores.

(c) What value is of particular interest to see whether or not it is in the interval?

(d) A student in your class wrote the following interpretation about the 95% confidence level used in making the interval. Is it a correct interpretation? If not, update it to make it correct.

"If this study were repeated many times, we would expect 95% of the resulting confidence intervals to contain the sample mean improvement in reasoning scores."

SPSS Note:

The differences were already computed and entered as the data. So to make a confidence interval with SPSS we would need to perform a one-sample t-test on the differences and specify a test value of 0. We should also check the options box to be sure the confidence level is the one you want, namely 95%.

One-Sample Test

	Test Value = 0					
					95% Confidence Interval of the Difference	
	t	df	Sig. (2-tailed)	Mean Difference	Lower	Upper
REASCORE	6.904	33	.000	3.62	2.55	4.68

If the before and after scores were entered into SPSS, then we would use a paired t-test option. SPSS would compute the differences for us and provide the confidence interval results. Details of the SPSS steps for analyzing paired data can be found in your Lab Workbook.

Below is the summary of confidence intervals for the big 5 parameters covered in Chapters 10 and 11. We have covered four of the five scenarios, those in the entire top half that are circled.

Population Proportion	Two Population Proportions	Population Mean
Parameter p	**Parameter** $p_1 - p_2$	**Parameter** μ
Statistic $\hat{p}$	**Statistic** $\hat{p}_1 - \hat{p}_2$	**Statistic** $\bar{x}$
Standard Error $\text{s.e.}(\hat{p}) = \sqrt{\frac{\hat{p}(1-\hat{p})}{n}}$	**Standard Error** $\text{s.e.}(\hat{p}_1 - \hat{p}_2) = \sqrt{\frac{\hat{p}_1(1-\hat{p}_1)}{n_1} + \frac{\hat{p}_2(1-\hat{p}_2)}{n_2}}$	**Standard Error** $\text{s.e.}(\bar{x}) = \frac{s}{\sqrt{n}}$
Confidence Interval $\hat{p} \pm z^* \text{s.e.}(\hat{p})$ **Conservative Conf. Interval** $\hat{p} \pm \frac{z^*}{2\sqrt{n}}$ **Sample Size** $n = \left(\frac{z^*}{2m}\right)^2$	**Confidence Interval** $(\hat{p}_1 - \hat{p}_2) \pm z^* \text{s.e.}(\hat{p}_1 - \hat{p}_2)$	**Confidence Interval** $\bar{x} \pm t^* \text{s.e.}(\bar{x})$ df = $n - 1$ **Paired Confidence Interval** $\bar{d} \pm t^* \text{s.e.}(\bar{d})$ df = $n - 1$

Two Population Means	
General	**Pooled**
Parameter $\mu_1 - \mu_2$	**Parameter** $\mu_1 - \mu_2$
Statistic $\bar{x}_1 - \bar{x}_2$	**Statistic** $\bar{x}_1 - \bar{x}_2$
Standard Error $\text{s.e.}(\bar{x}_1 - \bar{x}_2) = \sqrt{\frac{s_1^2}{n_1} + \frac{s_2^2}{n_2}}$	**Standard Error** pooled $\text{s.e.}(\bar{x}_1 - \bar{x}_2) = s_p\sqrt{\frac{1}{n_1} + \frac{1}{n_2}}$ where $s_p = \sqrt{\frac{(n_1-1)s_1^2 + (n_2-1)s_2^2}{n_1 + n_2 - 2}}$
Confidence Interval $(\bar{x}_1 - \bar{x}_2) \pm t^*(\text{s.e.}(\bar{x}_1 - \bar{x}_2))$ df = $\min(n_1 - 1, n_2 - 1)$	**Confidence Interval** $(\bar{x}_1 - \bar{x}_2) \pm t^*(\text{pooled s.e.}(\bar{x}_1 - \bar{x}_2))$ df = $n_1 + n_2 - 2$

Additional Notes

A place to ... jot down questions you may have and ask during office hours, take a few extra notes, write out an extra practice problem or summary completed in lecture, create your own short summary about this chapter.

Stat 250 Gunderson Lecture Notes
Learning about a Population Mean Difference

Part 3: Testing about a Population Mean of Paired Difference
Chapter 13: Section 3, HT Module 4

13.3 HT Module 4: Testing Hypotheses about the Population Mean of Paired Differences μ_d

An important special case of a single mean of a population occurs when two quantitative variables are **collected in pairs**, and we desire information about the **difference** between the two variables. For paired data designs, it is the ***differences*** that we are interested in analyzing. By focusing on the differences we again have just one sample of observations (the differences) and are able to perform a one-sample t-test on the differences.

The procedure for the significance test is really the same - the sample mean of the data, $\bar{x}$, is just written as $\bar{d}$, primarily to emphasize that this is a paired design with the differences being analyzed. The commonly used null hypothesis is that the population mean difference is $\mu_d = 0$.

Possible null and alternative hypotheses.

1. **H_0:** **versus H_a:**

2. **H_0:** **versus H_a:**

3. **H_0:** **versus H_a:**

Note: The format of the alternative hypothesis depends on the research question of interest and the order in which the differences were taken.

Test Statistic and Conditions for the Test

Test statistic = $\dfrac{\text{Sample statistic} - \text{Null value}}{\text{Standard error}}$

If H_0 is true, this test statistic has a ________________________ distribution.
We use this distribution to report the (bounds for the) *p*-value.

Conditions for the test: The difference is assumed to be normally distributed for the population (but if the sample size is large, this condition is less crucial). So you need to examine the differences graphically and assess if there are any extreme outliers or skewness in the differences. If so, either the sample size needs to be large or an alternative testing method may be required.

Page 508 of the textbook provides a nice summary of the steps for a paired *t*-test.

Try It! Knob Turning

A study involved n=25 right-handed students and a device with two different knobs (right-hand thread and left-hand thread). The **response** of interest is the time it takes to move knob indicator a fixed distance. The **question** of interest is to assess if right-hand threads are easier to turn on average. Use a 5% significance level.

a. Why is this a paired design and how should randomization be used in the experiment?

b. State the hypotheses. H_0: ________________ versus H_a: ________________

c. Perform the test.

Paired Samples Statistics

		Mean	N	Std. Deviation	Std. Error Mean
Pair 1	RTHREAD	104.00	25	15.93	3.19
	LTHREAD	117.44	25	27.26	5.45

Paired Samples Test

	Paired Differences					t	df	Sig. (2-tailed)
				90% Confidence Interval of the Difference				
	Mean	Std. Deviation	Std. Error Mean	Lower	Upper			
Pair 1 RTHREAD - LTHREAD	-13.44	23.06	4.61	-21.33	-5.55	-2.914	24	.008

d. Which are assumptions required for performing the paired t-test?

- the turning times for the right-hand threaded knob are independent of the turning times for the left-hand threaded knob.
- the turning times for the right hand threaded knob are normally distributed.
- the difference in turning times (diff = RT – LT) is normally distributed.

Stats 250 Formula Card Summary

We have covered confidence interval estimation and hypothesis testing for four of the five parameters, corresponding to the entire top half that is circled below.

Population Proportion	Two Population Proportions	Population Mean
Parameter p	**Parameter** $p_1 - p_2$	**Parameter** μ
Statistic $\hat{p}$	**Statistic** $\hat{p}_1 - \hat{p}_2$	**Statistic** $\bar{x}$
Standard Error $\text{s.e.}(\hat{p}) = \sqrt{\frac{\hat{p}(1-\hat{p})}{n}}$	**Standard Error** $\text{s.e.}(\hat{p}_1 - \hat{p}_2) = \sqrt{\frac{\hat{p}_1(1-\hat{p}_1)}{n_1} + \frac{\hat{p}_2(1-\hat{p}_2)}{n_2}}$	**Standard Error** $\text{s.e.}(\bar{x}) = \frac{s}{\sqrt{n}}$
Confidence Interval $\hat{p} \pm z^* \text{s.e.}(\hat{p})$ **Conservative Confidence Interval** $\hat{p} \pm \frac{z^*}{2\sqrt{n}}$	**Confidence Interval** $(\hat{p}_1 - \hat{p}_2) \pm z^* \text{s.e.}(\hat{p}_1 - \hat{p}_2)$	**Confidence Interval** $\bar{x} \pm t^* \text{s.e.}(\bar{x})$ df = $n - 1$ **Paired Confidence Interval** $\bar{d} \pm t^* \text{s.e.}(\bar{d})$ df = $n - 1$
Large-Sample *z*-Test $z = \frac{\hat{p} - p_0}{\sqrt{\frac{p_0(1-p_0)}{n}}}$ **Sample Size** $n = \left(\frac{z^*}{2m}\right)^2$	**Large-Sample *z*-Test** $z = \frac{\hat{p}_1 - \hat{p}_2}{\sqrt{\hat{p}(1-\hat{p})\left(\frac{1}{n_1} + \frac{1}{n_2}\right)}}$ Where $\hat{p} = \frac{n_1\hat{p}_1 + n_2\hat{p}_2}{n_1 + n_2}$	**One-Sample *t*-Test** $t = \frac{\bar{x} - \mu_0}{\text{s.e.}(\bar{x})} = \frac{\bar{x} - \mu_0}{s/\sqrt{n}}$ df = $n - 1$ **Paired *t*-Test** $t = \frac{\bar{d} - 0}{\text{s.e.}(\bar{d})} = \frac{\bar{d}}{s_d/\sqrt{n}}$ df = $n - 1$

Two Population Means	
General	**Pooled**
Parameter $\mu_1 - \mu_2$	**Parameter** $\mu_1 - \mu_2$
Statistic $\bar{x}_1 - \bar{x}_2$	**Statistic** $\bar{x}_1 - \bar{x}_2$
Standard Error $\text{s.e.}(\bar{x}_1 - \bar{x}_2) = \sqrt{\frac{s_1^2}{n_1} + \frac{s_2^2}{n_2}}$	**Standard Error** $\text{pooled s.e.}(\bar{x}_1 - \bar{x}_2) = s_p\sqrt{\frac{1}{n_1} + \frac{1}{n_2}}$ where $s_p = \sqrt{\frac{(n_1 - 1)s_1^2 + (n_2 - 1)s_2^2}{n_1 + n_2 - 2}}$
Confidence Interval $(\bar{x}_1 - \bar{x}_2) \pm t^*(\text{s.e.}(\bar{x}_1 - \bar{x}_2))$ df = $\min(n_1 - 1, n_2 - 1)$	**Confidence Interval** $(\bar{x}_1 - \bar{x}_2) \pm t^*(\text{pooled s.e.}(\bar{x}_1 - \bar{x}_2))$ df = $n_1 + n_2 - 2$
Two-Sample *t*-Test $t = \frac{\bar{x}_1 - \bar{x}_2 - 0}{\text{s.e.}(\bar{x}_1 - \bar{x}_2)} = \frac{\bar{x}_1 - \bar{x}_2}{\sqrt{\frac{s_1^2}{n_1} + \frac{s_2^2}{n_2}}}$ df = $\min(n_1 - 1, n_2 - 1)$	**Pooled Two-Sample *t*-Test** $t = \frac{\bar{x}_1 - \bar{x}_2 - 0}{\text{pooled s.e.}(\bar{x}_1 - \bar{x}_2)} = \frac{\bar{x}_1 - \bar{x}_2}{s_p\sqrt{\frac{1}{n_1} + \frac{1}{n_2}}}$ df = $n_1 + n_2 - 2$

Additional Notes

A place to ... jot down questions you may have and ask during office hours, take a few extra notes, write out an extra practice problem or summary completed in lecture, create your own short summary about this chapter.

Stat 250 Gunderson Lecture Notes
Learning about the Difference in Population Means

Part 1: Distribution for a Difference in Sample Means
Chapter 9: Section 8, SD Module 5

The Independent Samples Scenario

Recall that two samples are said to be **independent samples** when the measurements in one sample are not related to the measurements in the other sample. Independent samples are generated in a variety of ways.

Some common ways:

- **Random samples are taken separately from two populations** and the same response variable is recorded for each individual.
- **One random sample** is taken and a variable is recorded for each individual, but then **units are categorized as belonging to one population or another**, e.g. male/female.
- **Participants are randomly assigned to one of two treatment conditions**, such as diet or exercise, and the same response variable, such as weight loss, is recorded for each individual unit.

If the response variable is quantitative, a researcher might compare two independent groups by looking at the **difference between the two means**. Here is an example of an independent samples scenario from page 319 of your text.

Situation 5. Estimating the Difference Between Two Populations with Regard to the Mean of a Quantitative Variable

- *Example research questions:*

  How much difference is there in average weight loss for those who diet compared to those who exercise to lose weight?

  How much difference is there between the mean foot lengths of men and women?

- *Population parameter:* $\mu_1 - \mu_2$, where μ_1 and μ_2 represent the means in populations 1 and 2, respectively.
- *Sample estimate:* $\bar{x}_1 - \bar{x}_2$, the difference between the two sample means.

From Utts, Jessica M. and Robert F. Heckard. Mind on Statistics, Fourth Edition. 2012. Used with permission.

You can review page 338 for the short discussion and example regarding the difference in two population means.

9.8 SD Module 5: Sampling Distribution for the Difference in Two Sample Means

Who are the Speed Demons?
(Example 9.10 on pages 339-340 of the textbook)

A survey of college students included the question: "What is the fastest you have ever driven a car? _____ mph". Data for 87 males and 102 females responding to this question resulted in a mean speed of 107 mph for males and a mean speed of 88 mph for females.

A Typical Summary of Responses for a Two Independent Samples Problem:

Population	Sample Size	Sample Mean	Sample Standard Deviation
1 (males)	=87	=107	=17
2 (females)	=102	=88	s_2 =14

Let μ_1 be the population mean fastest speed for the male college student population.

Let μ_2 be the population mean fastest speed for the female college student population.

We want to learn about μ_1 and μ_2 and how they compare to each other. We could estimate the difference in population means $\mu_1 - \mu_2$ with the difference in the sample means $\bar{x}_1 - \bar{x}_2$. Will it be a good estimate?

Can anyone say how close this observed difference in sample means $\bar{x}_1 - \bar{x}_2$ of 19 mph is to the true difference in population means $\mu_1 - \mu_2$? _________

If we were to repeat this survey (with samples of the same sizes), would we get the same value for the difference in sample means? ___________________

Is a difference in the sample means of 19 mph large enough to convince us that there is a real difference in the means for the populations of students?

So what are the possible values for the difference in sample means $\bar{x}_1 - \bar{x}_2$ if we took many sets of independent random samples of the same sizes from these two populations? What would the distribution of the possible $\bar{x}_1 - \bar{x}_2$ values look like?

What can we say about the **distribution of the difference in two sample means**?

Using results from Section 8.8 (differences of independent random variables) and Section 9.6 (sampling distribution for a sample mean), the sampling distribution of the difference in two sample means $\bar{x}_1 - \bar{x}_2$ can be determined.

From Section 8.8 we learned that the mean of the difference is just the difference in the two means, and that the variance of the difference of two independent random variables is the sum of the variances.

From Section 9.6 we saw that the standard deviation of a sample mean is $\frac{\sigma}{\sqrt{n}}$.

We will apply these ideas to our newest parameter of interest, the difference in two sample means $\bar{x}_1 - \bar{x}_2$.

Sampling Distribution of the Difference in Two (Indep) Sample Means

If the two populations are normally distributed (or sample sizes are both large enough), then

$\bar{x}_1 - \bar{x}_2$ is (approximately)

Since the population standard deviations of σ_1 and σ_2 are generally not known, we will use the data to compute the standard error of the difference in sample means.

Standard Error of the Difference in Sample Means

$$\text{s.e.}(\bar{x}_1 - \bar{x}_2) = \sqrt{\frac{s_1^2}{n_1} + \frac{s_2^2}{n_2}}$$

where s_1 and s_2 are the two sample standard deviations

The standard error of $\bar{x}_1 - \bar{x}_2$ estimates, roughly, the average distance of the possible $\bar{x}_1 - \bar{x}_2$ values from $\mu_1 - \mu_2$. The possible $\bar{x}_1 - \bar{x}_2$ values result from considering all possible independent random samples of the same sizes from the same two populations.

Moreover, we can use this standard error to produce a range of values that we are very confident will contain the difference in the population means $\mu_1 - \mu_2$, namely, $\bar{x}_1 - \bar{x}_2 \pm$ (a few)s.e.($\bar{x}_1 - \bar{x}_2$). This is the basis for confidence interval for the difference in population means discussed in Chapter 11.

Looking ahead:

Do you think the 'few' in the above expression will be a z^* value or a t^* value?

What do you think will be the degrees of freedom?

We will use the standard error of the difference in the sample means to compute a standardized test statistic for testing hypotheses about the difference in the population means $\mu_1 - \mu_2$, namely,

$$\frac{\textbf{Sample statistic} - \textbf{Null value}}{\textbf{(Null) standard error}}$$

This is the basis for testing covered in Chapter 13.

Looking ahead:

Do you think the standardized test statistic will be a z statistic or a t statistic?

What do you think will be the most common null value used?

H_0: $\mu_1 - \mu_2$ = ____________

Additional Notes

A place to … jot down questions you may have and ask during office hours, take a few extra notes, write out an extra practice problem or summary completed in lecture, create your own short summary about this chapter.

Stat 250 Gunderson Lecture Notes
Learning about the Difference in Population Means

Part 2: Confidence Interval for a Difference in Population Means
Chapter 11: Section 4, CI Module 5

11.4 CI Module 5: Confidence Interval for the Difference in Two Population Means

Lesson 1: The General (Unpooled) Case

- We have two populations or groups from which independent samples are available, (or one population for which two groups can be formed using a categorical variable).
- The response variable is quantitative and we are interested in comparing the means for the two populations.

A Typical Summary of the Responses for a Two Independent Samples Problem:

Population	Sample Size	Sample Mean	Sample Standard Deviation
1	n_1	$\bar{x}_1$	s_1
2	n_2	$\bar{x}_2$	s_2

Let μ_1 be the mean response for the first population and μ_2 be the mean response for the second population.

Parameter of interest: the difference in the population means $\mu_1 - \mu_2$.

Sample estimate: the difference in the sample means $\bar{x}_1 - \bar{x}_2$.

Standard error: $\text{s.e.}(\bar{x}_1 - \bar{x}_2) = \sqrt{\frac{s_1^2}{n_1} + \frac{s_2^2}{n_2}}$

where s_1 and s_2 are the sample standard deviations.

So we have our estimate of the difference in the two population means, namely $\bar{x}_1 - \bar{x}_2$, and we have its standard error. To make our confidence interval, we need to know the multiplier. The **multiplier t^*** is a *t*-value such that the area between $-t^*$ and t^* equals the desired confidence level. The degrees of freedom for the *t*-distribution will depend on whether we use an *ugly* formula (used by software packages) or we use a conservative "by-hand" approach.

General **Two Independent-Samples *t* Confidence Interval for μ_1 - μ_2**

$$(\bar{x}_1 - \bar{x}_2) \pm t^*(\text{s.e.}(\bar{x}_1 - \bar{x}_2))$$

where $\text{s.e.}(\bar{x}_1 - \bar{x}_2) = \sqrt{\frac{s_1^2}{n_1} + \frac{s_2^2}{n_2}}$ and t^* is the appropriate value for a *t*-distribution, and the df can be found using an approximation or conservatively as df = smaller of ($n_1 - 1$ or $n_2 - 1$)

This interval requires we have independent random samples from normal populations.
If the sample sizes are large (both > 30), the assumption of normality is not so crucial and the result is approximate.

Lesson 2: The Pooled Case

If we can further ***assume the population variances are (unknown but) equal***, then there is a procedure for which the t^* multiplier is easier to find using an exact (not approximate) *t*-distribution. It involves pooling the sample variances for an overall estimate and updating the standard error accordingly.

It sometimes may be reasonable to assume that the measurements in the two populations have the same variances ...

so that ______________________ where _____ denotes the **common population variance**.

Since both sample variances would be estimating the common population variance, it would make sense to combine or ***pool*** **the two sample variances together** to form an overall estimate.

Pooled standard deviation: $s_p = \sqrt{\frac{(n_1 - 1)s_1^2 + (n_2 - 1)s_2^2}{n_1 + n_2 - 2}}$

Notes:

(1) Each sample variance is weighted by the corresponding degrees of freedom. So a larger sample size will result in a larger weight for that sample variance.

(2) The denominator gives the total degrees of freedom:
df =

Replacing the individual standard deviations s_1 and s_2 with the pooled version s_p in the formula for the standard error leads to the pooled standard error of $\bar{x}_1 - \bar{x}_2$ is given by:

Pooled s.e.($\bar{x}_1 - \bar{x}_2$) =

Pooled **Two Independent-Samples *t* Confidence Interval for μ_1 - μ_2**

$$(\bar{x}_1 - \bar{x}_2) \pm t^* (\text{pooled s.e.}(\bar{x}_1 - \bar{x}_2))$$

where $\text{pooled s.e.}(\bar{x}_1 - \bar{x}_2) = s_p \sqrt{\frac{1}{n_1} + \frac{1}{n_2}}$

and $s_p = \sqrt{\frac{(n_1 - 1)s_1^2 + (n_2 - 1)s_2^2}{n_1 + n_2 - 2}}$;

and t^* is the appropriate value for a $t(n_1 + n_2 - 2)$ distribution.

This interval requires we have independent random samples from normal populations with equal population variances.
If the sample sizes are large(both>30), the assumption of normality is not so crucial and the result is approximate.

Computer Package Notes:

(1) Most computer software provide results for both the unpooled and the pooled two independent samples *t* procedures.

(2) Some computer software also provide the results of a Levene's test for assessing if the population variances can be assumed equal.

The null hypothesis for this test is that the population variances are equal. So a small *p*-value for Levene's test would lead to rejecting that null hypothesis and concluding that the pooled procedure should not be used. Often a significance level of 10% is used for this condition checking. Your lab workbook provides more details about Levene's test.

(3) Another graphical tool for assessing equal population variance is side-by-side boxplots and comparing the IQRs. If the lengths of the boxes and overall ranges between the two groups are very different, the pooled method may not be reasonable.

Here are some **guidelines** to help assess which version to use (text pg 432 – 433).

- **If the sample standard deviations are similar**, the assumption of common population variance is reasonable and the pooled procedure can be used. If the sample sizes happen to be the same, the pooled and unpooled standard errors are equal anyway. The advantage for the pooled version is that finding the df is simpler.

  But the sample standard deviations are almost never exactly equal. So how similar is similar enough?

- **If the larger standard deviation is from the group with the larger sample size**, the pooled procedure is acceptable because it will be conservative (produce a wider interval).

- **If the smaller standard deviation is from the group with the larger sample size**, the pooled procedure can produce a misleading narrower interval.

Bottom-line: Pool if reasonable; but if the sample standard deviations are not similar, we have the unpooled procedure that can be used.

Try It! Comparing Reading Scores

A study compared data on reading scores for students taught by certified teachers versus those taught by uncertified teachers. Summaries of reading scores for both samples are provided.

Group Statistics

	GROUP	N	Mean	Std. Deviation	Std. Error Mean
READING SCORE	1	10	34.21	9.118	2.883
	2	10	28.20	7.036	2.225

Independent Samples Test

		Levene's Test for Equality of Variances		t-test for Equality of Means				
		F	Sig.	t	df	Sig. (2-tailed)	Mean Difference	Std. Error Difference
READING SCORE	Equal variances assumed	.104	.751	1.651	18	.116	6.01	3.642
	Equal variances not assumed			1.651	16.913	.117	6.01	3.642

All images

Let μ_1 denote the average reading score for the population of students taught by a certified teacher and μ_2 denote the average reading score for the population of students taught by an uncertified teacher.

a. One of the assumptions for the pooled two independent samples confidence interval to be valid is that the 2 populations (from which we took our samples) have the same standard deviation. Look at the two sample standard deviations and the Levene's test result. Does the assumption seem to hold (at the 10% level)? Explain.

b. Give a 95% confidence interval for the difference in the population mean reading scores, that is, for $\mu_1 - \mu_2$.

c. Based on the interval, does there appear to be a difference in the mean reading scores for the two populations? Explain.

Try It! Stroop's Word Color Test

In Stroop's Word Color Test, 100 words that are color names are shown in colors different from the word. For example, the word red might be displayed in blue. The task is to correctly identify the display color of each word; in the example just given the correct response would be blue.

Gustafson and Kallmen (1990) recorded the time needed to complete this test for $n = 16$ individuals after they had consumed alcohol and for $n = 16$ other individuals after they had consumed a placebo drink flavored to taste as if it contained alcohol. Each group was balanced with 8 men and 8 women.

In the alcohol group, the mean completion time was 113.75 seconds and standard deviation was 22.64 seconds. In the placebo group, the mean completion time was 99.87 seconds and standard deviation was 12.04 seconds.

Group	*Sample size*	*Sample mean*	*Sample standard deviation*
1 = alcohol	16	113.75	22.64
2 = placebo	16	99.87	12.04

We can assume that the two samples are independent random samples, that the model for completion time is normal for each population.

a. What graph(s) would you make to check the normality condition? Be specific.

b. How did the two groups compare descriptively?

c. Which procedure? Pooled or unpooled? Why?

d. Calculate a **95% confidence interval for the difference in population means**.

e. Based on the confidence interval, can we conclude that the population means for the two groups are different? Why or why not?

What if?

Suppose the researchers Gustafson and Kallmen were convinced (based on past results) that the underlying population variances were equal, so they prefer that a pooled confidence interval be constructed.

The estimate of the common population standard deviation would be:

$$s_p = \sqrt{\frac{(n_1-1)s_1^2+(n_2-1)s_2^2}{n_1+n_2-2}} = \sqrt{\frac{(16-1)(22.64)^2+(16-1)(12.04)^2}{16+16-2}} = \sqrt{328.77} = 18.13$$

The pooled standard error for the difference in the two sample means would be:

$$\text{pooled s.e.}(\bar{x}_1-\bar{x}_2) = s_p\sqrt{\frac{1}{n_1}+\frac{1}{n_2}} = 18.13\sqrt{\frac{1}{16}+\frac{1}{16}} = 6.41$$

which is the same as the unpooled standard error since the sample sizes were equal. The t^* multiplier would be based on df = 16 + 16 – 2 = 30, so t^* = 2.04 (from Table A.2).

The 95% Pooled Confidence Interval would be:

$$(\bar{x}_1-\bar{x}_2) \pm t^*(\text{pooled s.e.}(\bar{x}_1-\bar{x}_2))$$

$\rightarrow (13.88) \pm (2.04)(6.41) \rightarrow 13.88 \pm 13.08 \rightarrow (0.80, 26.96)$

This interval still does not include 0, so the same decision would be made; however, the interval is a bit narrower.

In this example, the unpooled interval may be a bit conservative (wider) but the evidence is still strong to state the two population means appear to differ.

Note: Computer packages will compute both intervals for us, so we would not have to do all of the computations above by hand!

Stat 250 Formula Card

Here is the summary of the confidence intervals for the big 5 parameters covered in Chapters 10 and 11. We have now covered all 5 scenarios.

Population Proportion	**Two Population Proportions**	**Population Mean**
Parameter p	**Parameter** $p_1 - p_2$	**Parameter** μ
Statistic $\hat{p}$	**Statistic** $\hat{p}_1 - \hat{p}_2$	**Statistic** $\bar{x}$
Standard Error $\text{s.e.}(\hat{p}) = \sqrt{\frac{\hat{p}(1-\hat{p})}{n}}$	**Standard Error** $\text{s.e.}(\hat{p}_1 - \hat{p}_2) = \sqrt{\frac{\hat{p}_1(1-\hat{p}_1)}{n_1} + \frac{\hat{p}_2(1-\hat{p}_2)}{n_2}}$	**Standard Error** $\text{s.e.}(\bar{x}) = \frac{s}{\sqrt{n}}$
Confidence Interval $\hat{p} \pm z^* \text{s.e.}(\hat{p})$ **Conservative Conf. Interval** $\hat{p} \pm \frac{z^*}{2\sqrt{n}}$ **Sample Size** $n = \left(\frac{z^*}{2m}\right)^2$	**Confidence Interval** $(\hat{p}_1 - \hat{p}_2) \pm z^* \text{s.e.}(\hat{p}_1 - \hat{p}_2)$	**Confidence Interval** $\bar{x} \pm t^* \text{s.e.}(\bar{x})$ df = $n - 1$ **Paired Confidence Interval** $\bar{d} \pm t^* \text{s.e.}(\bar{d})$ df = $n - 1$

Two Population Means	
General	**Pooled**
Parameter $\mu_1 - \mu_2$	**Parameter** $\mu_1 - \mu_2$
Statistic $\bar{x}_1 - \bar{x}_2$	**Statistic** $\bar{x}_1 - \bar{x}_2$
Standard Error $\text{s.e.}(\bar{x}_1 - \bar{x}_2) = \sqrt{\frac{s_1^2}{n_1} + \frac{s_2^2}{n_2}}$	**Standard Error** $\text{pooled s.e.}(\bar{x}_1 - \bar{x}_2) = s_p \sqrt{\frac{1}{n_1} + \frac{1}{n_2}}$ where $s_p = \sqrt{\frac{(n_1 - 1)s_1^2 + (n_2 - 1)s_2^2}{n_1 + n_2 - 2}}$
Confidence Interval $(\bar{x}_1 - \bar{x}_2) \pm t^* (\text{s.e.}(\bar{x}_1 - \bar{x}_2))$ df = $\min(n_1 - 1, n_2 - 1)$	**Confidence Interval** $(\bar{x}_1 - \bar{x}_2) \pm t^* (\text{pooled s.e.}(\bar{x}_1 - \bar{x}_2))$ df = $n_1 + n_2 - 2$

Additional Notes

A place to ... jot down questions you may have and ask during office hours, take a few extra notes, write out an extra practice problem or summary completed in lecture, create your own short summary about this chapter.

Stat 250 Gunderson Lecture Notes
Learning about the Difference in Population Means

Part 3: Testing about a Difference in Population Means
Chapter 13: Section 4, HT Module 5; Section 13.6

13.4 HT Module 5: Testing Hypotheses about the Difference in Two Population Means

- We have two populations or groups from which independent samples are available, (or one population for which two groups can be formed using a categorical variable).
- The response variable is quantitative and we are interested in testing hypotheses about the means for the two populations.

A Typical Summary of the Responses for a Two Independent Samples Problem:

Population	Sample Size	Sample Mean	Sample Standard Deviation
1	n_1	$\bar{x}_1$	s_1
2	n_2	$\bar{x}_2$	s_2

Let μ_1 be the mean response for the first population and μ_2 be the mean response for the second population.

Parameter of interest: the difference in the population means $\mu_1 - \mu_2$.

Sample estimate: the difference in the sample means $\bar{x}_1 - \bar{x}_2$.

Standard error: $$\text{s.e.}(\bar{x}_1 - \bar{x}_2) = \sqrt{\frac{s_1^2}{n_1} + \frac{s_2^2}{n_2}}$$

where s_1 and s_2 are the two sample standard deviations

Pooled standard error: $$\text{pooled s.e.}(\bar{x}_1 - \bar{x}_2) = s_p\sqrt{\frac{1}{n_1} + \frac{1}{n_2}}$$

where $$s_p = \sqrt{\frac{(n_1-1)s_1^2 + (n_2-1)s_2^2}{n_1 + n_2 - 2}}$$

Recall there are two methods for conducting inference for the difference between two population means for independent samples – the **General (Unpooled) Case** and the **Pooled Case**. Both require we have independent random samples from normal populations (but if the sample sizes are large, the assumption of normality is not so crucial). Both will result in a t-test statistic, but the standard error used in the denominator differ as well as the degrees of freedom used for computing the *p*-value using a t-distribution.

Here is the summary for these two significance tests:

Possible null and alternative hypotheses.

1. $\mathbf{H_0}$: **versus** $\mathbf{H_a}$:

2. $\mathbf{H_0}$: **versus** $\mathbf{H_a}$:

3. $\mathbf{H_0}$: **versus** $\mathbf{H_a}$:

$$\textbf{Test statistic} = \frac{\textbf{Sample statistic} - \textbf{Null value}}{\textbf{Standard error}}$$

General	Pooled
(Unpooled) Two-Sample *t*-Test $$t = \frac{\bar{x}_1 - \bar{x}_2 - 0}{\text{s.e.}(\bar{x}_1 - \bar{x}_2)} = \frac{\bar{x}_1 - \bar{x}_2}{\sqrt{\frac{s_1^2}{n_1} + \frac{s_2^2}{n_2}}}$$ where df = $\min(n_1 - 1, n_2 - 1)$	**Pooled Two-Sample *t*-Test** $$t = \frac{\bar{x}_1 - \bar{x}_2 - 0}{\text{pooled s.e.}(\bar{x}_1 - \bar{x}_2)} = \frac{\bar{x}_1 - \bar{x}_2}{s_p\sqrt{\frac{1}{n_1} + \frac{1}{n_2}}}$$ where $s_p = \sqrt{\frac{(n_1 - 1)s_1^2 + (n_2 - 1)s_2^2}{n_1 + n_2 - 2}}$ and df = $n_1 + n_2 - 2$

Recall the **guidelines** to assess which version to use: (text pages 518 – 519)

- **If the sample standard deviations are similar**, the assumption of common population variance is reasonable and the pooled procedure can be used. If the sample sizes happen to be the same, the pooled and unpooled standard errors are equal anyway. The advantage for the pooled version is that finding the df is simpler.
- **If the larger standard deviation is from the group with the larger sample size**, the pooled procedure is acceptable because it will be conservative (produce a wider interval).
- **If the smaller standard deviation is from the group with the larger sample size**, the pooled procedure can produce a misleading narrower interval.

Bottom-line: Pool if reasonable; but if the sample standard deviations are not similar, we have the unpooled procedure that can be used.

Try It! Effect of Beta-blockers on pulse rate

Do beta-blockers reduce the pulse rate? In a study of heart surgery, 60 subjects were randomly divided into two groups of 30. One group received a beta-blocker while the other group was given a placebo. The pulse rate at a particular time during the surgery was measured. The results are given below.

Group	*Sample size*	*Sample mean*	*Sample standard deviation*
1=beta-blockers	30	65.2	7.8
2=placebo	30	70.3	8.4

a. State the hypotheses to assess if beta-blockers reduce pulse rate on average.

   H_0: ____________________ versus H_a: ________________________

b. Which test will you perform – the pooled or unpooled test? Explain.

c. Perform the t-test. Show all steps. Are the results significant at a 5% level?

Try It! Does the Drug Speed Learning?

In an animal-learning experiment, a researcher wanted to assess if a particular drug ***speeds*** learning. One group of 5 rats (Group 1 = control) is required to learn to run a maze without use of the drug, whereas a second independent group of 8 rats (Group 2 = experimental) is administered the drug. The running times (time to complete the maze) for the rats in each group were entered into SPSS.

Group Statistics

	Group	N	Mean	Std. Deviation	Std. Error Mean
Time to complete maze	Control	5	46.80	3.42	1.53
	Experimental	8	38.38	4.78	1.69

Independent Samples Test

		Levene's Test for Equality of Variances		t-test for Equality of Means				
		F	Sig.	t	df	Sig. (2-tailed)	Mean Difference	Std. Error Difference
Run Time	Equal variances assumed	1.09	.32	3.41	11	.006	8.42	2.47
	Equal variances not assumed			3.70	10.653	.004	8.42	2.28

Conduct the test to address the theory of the researcher (state the null and alternate hypotheses, report the test statistic, *p*-value, and state your decision and conclusion at the 5% level of significance).

H_0: ______________________

H_a: ______________________

Test statistic: ____________________________

p-value: ________________________________

Decision: (circle one) **Fail to reject H_0** **Reject H_0**

Thus ...

Try It! Eat that Dark Chocolate

An Ann Arbor News article entitled: Dark Chocolate may help blood flow, reported the results of a study in which researchers fed a small 1.6-ounce bar of dark chocolate to each of 22 volunteers daily for two weeks. Half of the subjects were randomly selected and assigned to receive bars containing dark chocolate's typically high levels of flavonoids, and the other half received placebo bars with just trace amounts of flavonoids. The ability of the brachial artery to dilate *significantly* improved for those in the high-flavonoid group compared to those in the placebo group.

Let μ_1 represent the population average improvement in blood flow for those on the high-flavonoid diet and μ_2 represent the population average improvement in blood flow for those on the placebo diet. The researchers tested that the high-flavonoid group would have a higher average improvement in blood flow.

a. State the null and alternate hypotheses

   H_0: ____________________ versus H_a: ________________________

b. The researchers conducted a pooled two sample t-test. The two assumptions about the data are that the two samples are independent random samples.

   i. Clearly state one of the remaining two assumptions regarding the populations that are required for this test to be valid.

   ii. Explain how you would use the data to assess if the above assumption in part (i) is reasonable. (Be specific.)

c. A significance level of 0.05 was used. Based on the statements reported above, what can you say about the *p*-value? Clearly circle your answer:

   ***p*-value > 0.05** ***p*-value ≤ 0.05** **can't tell**

d. The researchers also found that concentrations of the cocoa flavonoid epicatechin soared in blood samples taken from the group that received the high-flavonoid chocolate, rising from a baseline of25.6 nmol/L to 204.4 nmol/L. In the group that received the low-flavonoid chocolate, concentrations of epicatechin decreased slightly, from a baseline of 17.9 nmol/L to 17.5 nmol/L. The average improvement for the high-flavonoid group of 204.4 – 25.6 = 178.8 nmol/L is a ... (circle all correct answers):

   parameter **statistic** **sample mean** **population mean** **sampling distribution**

13.6 Choosing an Appropriate Inference Procedure

Now that we have covered the inference techniques for all five of the big five parameters, review this section (pages 523-524 of the textbook) which provides some guidance on how to identify the appropriate procedure based on the research question(s) of interest.

Questions to Ask:

- **Is the response variable measured quantitative or categorical?**
  - Categorical → Proportions, percentages
    - p: One population proportion
    - $p_1 - p_2$: Difference between two population proportions
  - Quantitative → Means
    - μ: One population mean
    - μ_d: Paired difference population mean
    - $\mu_1 - \mu_2$: Difference between two population means

- **How many samples?**
  - If two sets of measurements – are they paired or independent?

- **What is the main purpose?**
  - To estimate a numerical value of a parameter? → confidence interval
  - To make a 'maybe not' or 'maybe yes' type of conclusion about a specific hypothesized value? → hypothesis test

Table 13.3 page 524

Variable Type (Parameter Type)	One Sample (No Pairing)	Paired Data	Two Independent Samples
Categorical (proportions)	p	None (at this time)	$p_1 - p_2$
Quantitative (means)	μ	μ_d	$\mu_1 - \mu_2$

Hypothesis Testing Procedures

Summary Table from page 534

IN SUMMARY Hypothesis Testing Procedures

The basic structure of hypothesis testing for the parameters in the table below is as follows:

1. Null hypothesis: H_0: parameter = null value

   Alternative hypothesis: $\left\{\begin{array}{l} H_a\text{: parameter} \neq \text{null value} \\ H_a\text{: parameter} > \text{null value} \\ H_a\text{: parameter} < \text{null value} \end{array}\right\}$ Choose one.

2. Test statistic: t or $z = \dfrac{\text{Sample statistic} - \text{Null value}}{\text{Null standard error}}$

3. p-value: For H_a: parameter ≠ null value, p-value is 2 × area above $|t|$ or $|z|$.
   For H_a: parameter > null value, p-value is the area *above* t or z, even if t or z is negative.
   For H_a: parameter < null value, p-value is the area *below* t or z, even if t or z is positive.

4. Decision: If p-value > significance level, do not reject H_0.
   If p-value ≤ significance level, reject H_0, accept H_a, and conclude that the result is statistically significant.

5. Conclusion: Report the conclusion in the context of the situation.

	Parameter	Statistic	Null Value	Null Standard Error	t or z; df for t	Example
One Proportion	p	$\hat{p}$	p_0	$\sqrt{\dfrac{p_0(1-p_0)}{n}}$	z	12.12
Difference between Proportions	$p_1 - p_2$	$\hat{p}_1 - \hat{p}_2$	Usually 0	$\sqrt{\hat{p}(1-\hat{p})\left(\dfrac{1}{n_1}+\dfrac{1}{n_2}\right)}$	z	12.17
One Mean	μ	$\bar{x}$	μ_0	$\dfrac{s}{\sqrt{n}}$	t df = $n - 1$	13.1
Mean Difference, Paired Data	μ_d	$\bar{d}$	Usually 0	$\dfrac{s_d}{\sqrt{n}}$	t df = $n - 1$	13.2
Difference between Means (unpooled)	$\mu_1 - \mu_2$	$\bar{x}_1 - \bar{x}_2$	Usually 0	$\sqrt{\dfrac{s_1^2}{n_1}+\dfrac{s_2^2}{n_2}}$	t df = Welch's approximation or smaller of $n_1 - 1$, $n_2 - 1$	13.4
Difference between Means (pooled)	$\mu_1 - \mu_2$	$\bar{x}_1 - \bar{x}_2$	Usually 0	$s_p\sqrt{\dfrac{1}{n_1}+\dfrac{1}{n_2}}$	t df = $n_1 + n_2 - 2$	13.7

Inference Procedures for the Big 5 Parameters

Stats 250 Formula Card Summary

Population Proportion	Two Population Proportions	Population Mean
Parameter	**Parameter** $p_1 - p_2$	**Parameter** μ
Statistic $\hat{p}$	**Statistic** $\hat{p}_1 - \hat{p}_2$	**Statistic** $\bar{x}$
Standard Error $\text{s.e.}(\hat{p}) = \sqrt{\frac{\hat{p}(1-\hat{p})}{n}}$	**Standard Error** $\text{s.e.}(\hat{p}_1 - \hat{p}_2) = \sqrt{\frac{\hat{p}_1(1-\hat{p}_1)}{n_1} + \frac{\hat{p}_2(1-\hat{p}_2)}{n_2}}$	**Standard Error** $\text{s.e.}(\bar{x}) = \frac{s}{\sqrt{n}}$
Confidence Interval $\hat{p} \pm z^* \text{s.e.}(\hat{p})$ **Conservative Confidence Interval** $\hat{p} \pm \frac{z^*}{2\sqrt{n}}$	**Confidence Interval** $(\hat{p}_1 - \hat{p}_2) \pm z^* \text{s.e.}(\hat{p}_1 - \hat{p}_2)$	**Confidence Interval** $\bar{x} \pm t^* \text{s.e.}(\bar{x})$ df = $n - 1$ **Paired Confidence Interval** $\bar{d} \pm t^* \text{s.e.}(\bar{d})$ df = $n - 1$
Large-Sample *z*-Test $z = \frac{\hat{p} - p_0}{\sqrt{\frac{p_0(1-p_0)}{n}}}$ **Sample Size** $n = \left(\frac{z^*}{2m}\right)^2$	**Large-Sample *z*-Test** $z = \frac{\hat{p}_1 - \hat{p}_2}{\sqrt{\hat{p}(1-\hat{p})\left(\frac{1}{n_1} + \frac{1}{n_2}\right)}}$ Where $\hat{p} = \frac{n_1\hat{p}_1 + n_2\hat{p}_2}{n_1 + n_2}$	**One-Sample *t*-Test** $t = \frac{\bar{x} - \mu_0}{\text{s.e.}(\bar{x})} = \frac{\bar{x} - \mu_0}{s/\sqrt{n}}$ df = $n - 1$ **Paired *t*-Test** $t = \frac{\bar{d} - 0}{\text{s.e.}(\bar{d})} = \frac{\bar{d}}{s_d/\sqrt{n}}$ df = $n - 1$

Two Population Means	
General	**Pooled**
Parameter $\mu_1 - \mu_2$	**Parameter** $\mu_1 - \mu_2$
Statistic $\bar{x}_1 - \bar{x}_2$	**Statistic** $\bar{x}_1 - \bar{x}_2$
Standard Error $\text{s.e.}(\bar{x}_1 - \bar{x}_2) = \sqrt{\frac{s_1^2}{n_1} + \frac{s_2^2}{n_2}}$	**Standard Error** $\text{pooled s.e.}(\bar{x}_1 - \bar{x}_2) = s_p\sqrt{\frac{1}{n_1} + \frac{1}{n_2}}$ where $s_p = \sqrt{\frac{(n_1 - 1)s_1^2 + (n_2 - 1)s_2^2}{n_1 + n_2 - 2}}$
Confidence Interval $(\bar{x}_1 - \bar{x}_2) \pm t^*(\text{s.e.}(\bar{x}_1 - \bar{x}_2))$ df = $\min(n_1 - 1, n_2 - 1)$	**Confidence Interval** $(\bar{x}_1 - \bar{x}_2) \pm t^*(\text{pooled s.e.}(\bar{x}_1 - \bar{x}_2))$ df = $n_1 + n_2 - 2$
Two-Sample *t*-Test $t = \frac{\bar{x}_1 - \bar{x}_2 - 0}{\text{s.e.}(\bar{x}_1 - \bar{x}_2)} = \frac{\bar{x}_1 - \bar{x}_2}{\sqrt{\frac{s_1^2}{n_1} + \frac{s_2^2}{n_2}}}$ df = $\min(n_1 - 1, n_2 - 1)$	**Pooled Two-Sample *t*-Test** $t = \frac{\bar{x}_1 - \bar{x}_2 - 0}{\text{pooled s.e.}(\bar{x}_1 - \bar{x}_2)} = \frac{\bar{x}_1 - \bar{x}_2}{s_p\sqrt{\frac{1}{n_1} + \frac{1}{n_2}}}$ df = $n_1 + n_2 - 2$

Additional Notes

A place to ... jot down questions you may have and ask during office hours, take a few extra notes, write out an extra practice problem or summary completed in lecture, create your own short summary about this chapter.

Stat 250 Gunderson Lecture Notes
Chapter 16: Analysis of Variance
Sections 1, 2, and 3

Data! Data! Data! I can't make bricks without clay! -- *Sherlock Holmes*

We have already been introduced to the concept of comparing the means of two populations when the data gathered represent independent random samples from normal populations. When the response was quantitative, we learned about two methods, an unpooled method and a pooled method.

We turn to Chapter 16 to discuss a method that allows us to compare the means of two or more normal populations based on independent random samples when the population variances are assumed to be equal. This method is called "**ANALYSIS OF VARIANCE**" (abbreviated **ANOVA**) and is **an extension of the two independent samples POOLED t-test from Chapter 13.**

Let's step back for a moment to our two independent samples *t*-test. The purpose of this test was to decide whether or not two population means were equal:

H_0: ______________________

The test was based on a *t* statistic that had ______________degrees of freedom.

One-way ANOVA is basically an extension of our two independent samples *t*-test to handling more than 2 populations. One-way ANOVA is a technique for testing whether or not the means of several populations are equal.

Picture:

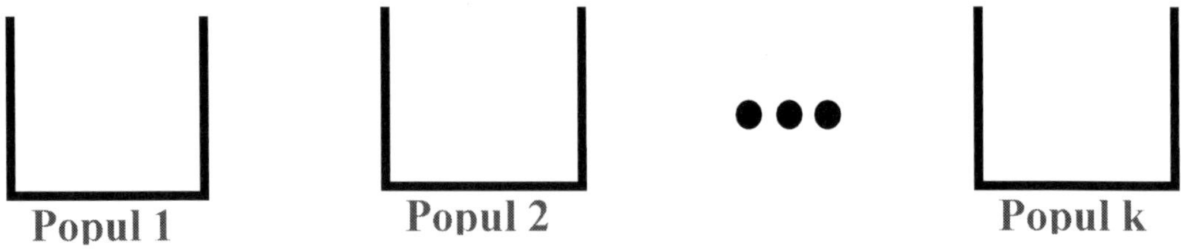

The assumptions are an extension of those for the two independent samples *t*-test to *k* groups.

- Each sample is a ... **random sample**
- The *k* random samples are ... **independent**
- For each of population the model for the response is... **a normal distribution**
- The *k* population variances are .... **equal**

The ANOVA Hypotheses:

H_0: ______________________ versus H_a: ____________________________________

Notice that this alternative does not require that all the population means be different from each other.

One possible H_a picture

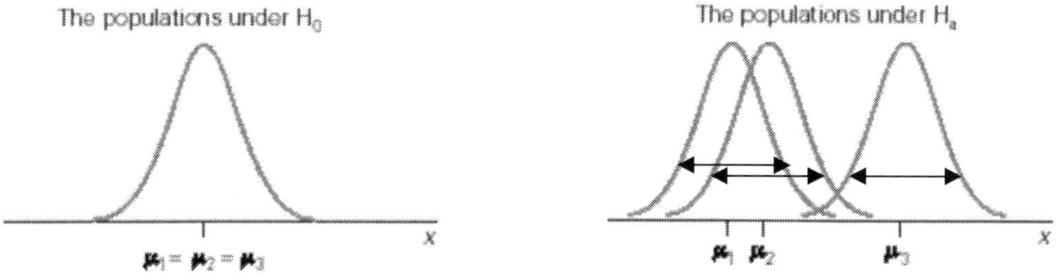

Question: What call a technique for testing the **equality of the means** " analysis of **VARIANCE**"?

Answer: We are going to **compare two estimators of the common population variance,** σ^2.

- MS Groups (Mean Square between the Groups):

- MSE (Mean Square Within or due to Error):

These two estimates are used to form the *F* statistic:

$$F = \frac{\text{Variation among sample means}}{\text{Natural variation within groups}} = \frac{MS\text{ Groups}}{MSE}.$$

If this *F* ratio is too ____________________ we would reject the null hypothesis.

The Logic behind the ANOVA F-Test

Look at the plots below. For each Scenario, we have plotted data obtained by taking independent random samples of size 10 from three populations.

For Scenarios A and B, the three populations each had a normal distribution and the population means were 60, 65, and 70, respectively. So the population means are indeed not all equal.

In Scenario A, the population standard deviations were all equal to 1.5. In Scenario B, the population standard deviations were all equal to 3. So in each case the assumption that the populations have equal standard deviations is met.

Scenario A

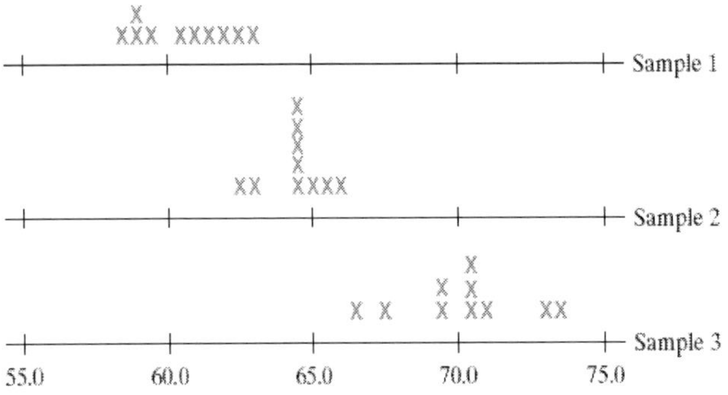

- Samples from 3 populations whose **means are different**.
- **Variability within** each population is **small**.
- **Difference between sample means more readily seen.**
- **F** statistic somewhat **big**.

Scenario B

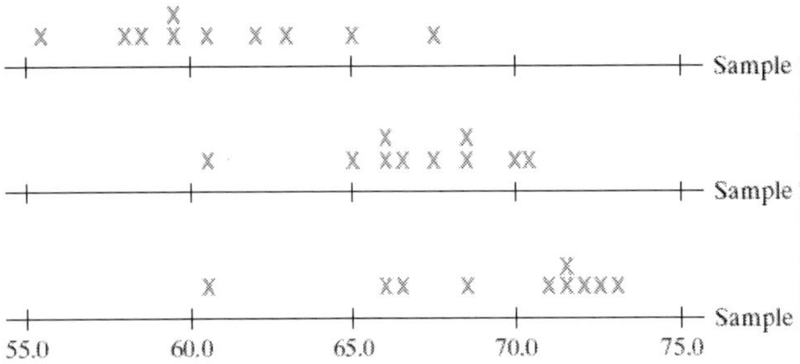

- Samples from 3 populations whose **means are different**.
- **Variability within** each population is **larger**.
- **Difference between sample means not readily seen.**
- **F** statistic **smaller**.

Which of the above two scenarios do you think would provide more evidence that at least one of the population means is different from the others? Scenario A or Scenario B?

Below is a final set of plots for three independent random samples of size 10 each taken from a population with a normal model with a population mean of 65 and population standard deviation of 1.5. So in Scenario C, the population means are indeed all equal—that is, the null hypothesis tested in one-way ANOVA is true. Notice that, although the population means were all equal, there is still some of variation between the sample means.

Also in Scenario C there is still some natural variation within the samples, making the slight variation between the sample means hardly noticeable. The data in Scenario C do not provide evidence that the population means are different.

Scenario C

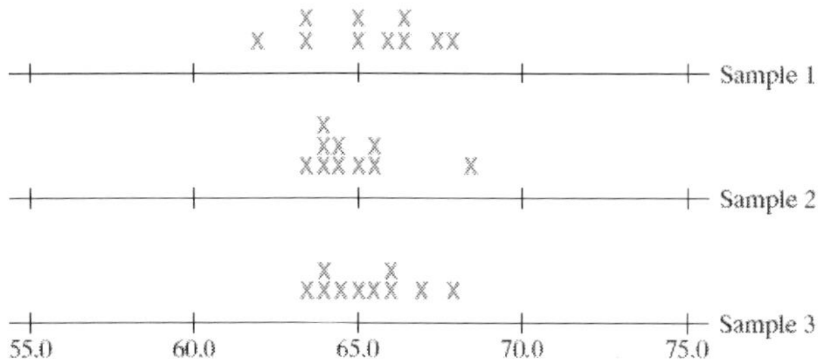

- Samples from 3 populations whose **means are all equal**.
- Still **some variability within** each population.
- **Very little differences between the sample means.**
- F statistic very **small**.

The F-statistic will be sensitive to differences between the sample means. The larger the **variation between the sample means**, the larger the value of the F-statistic and larger values of the F-statistic provide more support for rejecting the null hypothesis. The variation between the sample means was greatest for Scenarios A and B compared to Scenario C. The **natural variation within the samples** was greatest for Scenario B compared to Scenarios A and C. The F-statistic is the ratio of these two measures of variation:

$$F = \frac{\text{Variation among sample means}}{\text{Natural variation within groups}}$$

So which scenario would you expect to result in the largest value of the F-statistic? Provided below are the values of the F-statistic for the test of equality of the population means.

Scenario	Value of F Statistic	*p*-value
A (H_a is true)	F = 80.4	0.0000
B (H_a is true)	F = 16.4	0.01
C (H_0 is true)	F = 0.17	0.84

Note the value of the F-statistic is smallest and the *p*-value the largest when the null hypothesis is true (Scenario C). For Scenarios A and B, the population means are different, but the smaller population standard deviation in Scenario A accentuates the differences by producing a larger F-ratio and an extremely small *p*-value. **A larger F-statistic value (and thus smaller *p*-value) corresponds to more evidence that the population means are not all equal.**

Computing the F Test Statistic

We will see how to get MS Groups and MSE and perform the *F* test. These two mean squares will be a sum of squares (SS) divided by a corresponding degrees of freedom (DF).

The data can be generically represented below, where $X_{ij} = j^{th}$ observation from the ith population. However we really don't have to worry too much about these subscripts, as we will go through the steps using words!

Data from Population1	**Data from Population 2**	**...**	**Data from Population *k***
X_{11}	X_{21}		X_{k1}
X_{12}	X_{22}		X_{k2}
$\vdots$	$\vdots$		$\vdots$
X_{1n_1}	X_{2n_2}		X_{kn_k}

The details leading to the *F* statistic are presented in six steps, ending with an ANOVA table.

Step 1: Calculate the mean and variance for each sample: $\bar{x}_i,\ s_i^2$

Step 2: Calculate the overall sample mean
(using all $N = n_1 + n_2 + \ldots + n_k$ observations): $\bar{x}$

Step 3: Calculate the sum of squares between groups:

$$\text{SS Groups} = \sum_{groups} n_i \left(\bar{x}_i - \bar{x}\right)^2$$

Step 4: Calculate the sum of squares within groups (due to error):

$$\text{SSE} = \sum_{groups} \left(n_i - 1\right) s_i^2$$

Step 5: OPTIONAL: Calculate the total sum of squares:

$$\text{SS Total} = \sum_{values} \left(x_{ij} - \bar{x}\right)^2$$

Step 6: Fill in the ANOVA table:

Source	Sum of Squares	DF	Mean Square	F
Groups	SS Groups			
Error (Within)	SSE			
Total	SS Total			

If H_0 is true, then the *F* statistic, $F = \frac{MS\text{ Groups}}{MSE}$, has an $F(k - 1, N - k)$ distribution. Below are a few pictures of some F distributions.

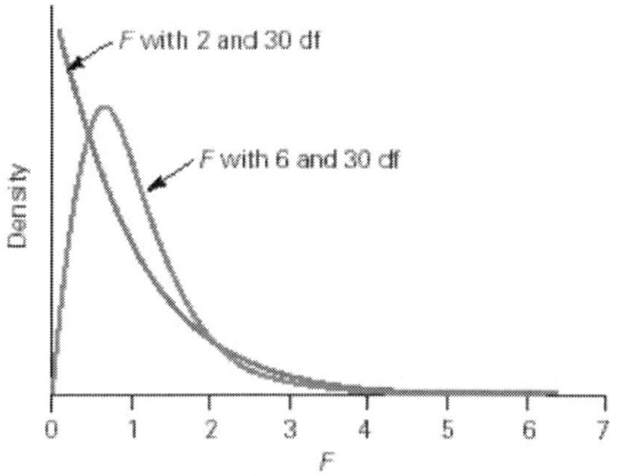

From Utts, Jessica M. and Robert F. Heckard. Mind on Statistics, Fourth Edition. 2012. Used with permission.

Table A.4 provides percentiles of an *F* distribution. However, standard computer output also provides the exact *p*-value and completed ANOVA table. We will rely on SPSS output to provide the *p*-value, but you should know how the ANOVA table is constructed and be able to sketch a picture of the *p*-value for an F-test.

Stat 250 Formula Card Summary of ANOVA:

One-Way ANOVA		
SS Groups = $SSG = \sum_{\text{groups}} n_i(\bar{x}_i - \bar{x})^2$	MS Groups = MSG = $\frac{SSG}{k-1}$	**ANOVA Table**
SS Error = $SSE = \sum_{\text{groups}} (n_i - 1)s_i^2$	MS Error = $MSE = s_p^2 = \frac{SSE}{N-k}$	**Source** / **SS** / **DF** / **MS** **Groups** / SS Groups / $k-1$ / MS Groups **Error** / SS Error / $N-k$ / MS Error **Total** / SSTO / $N-1$
SS Total = $SSTO = \sum_{\text{values}} (x_{ij} - \bar{x})^2$	$F = \frac{\text{MS Groups}}{\text{MS Error}}$	
Confidence Interval	$\bar{x}_i \pm t^* \frac{s_p}{\sqrt{n_i}}$ df = $N - k$	Under H_0, the *F* statistic follows an $F(k-1, N-k)$ distribution.

Try It! Comparing 3 Drugs

We wish to compare three drugs for treating some disease. A quantitative response is measured such that a smaller value indicates a more favorable response. A total of $N = 19$ patients are randomly assigned to one of the three drug (treatment) groups. The data are provided below:

Data from Drug 1	Data from Drug 2	Data from Drug 3
7.3	7.1	5.8
8.2	10.6	6.5
10.1	11.2	8.8
6.0	9.0	4.9
9.5	8.5	7.9
	10.9	8.5
	7.8	5.2

Let's first look at this data graphically.

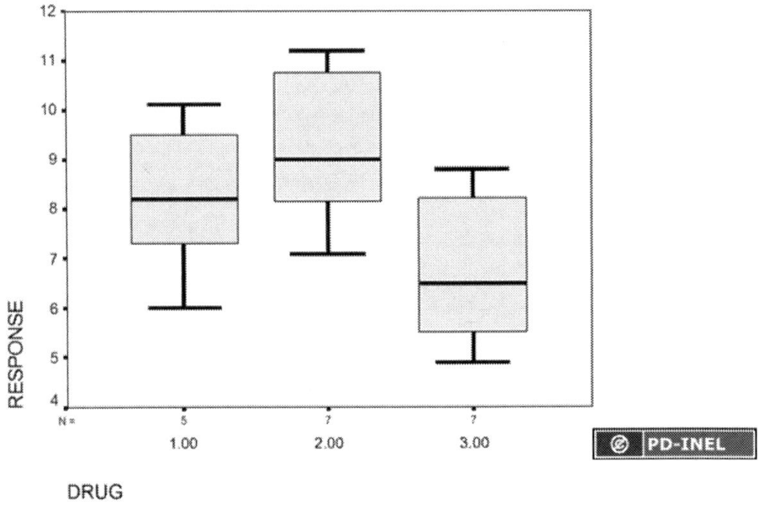

Recall the assumptions for performing an *F*-test and think about how you would check these assumptions.

- Each sample is a ... **random sample**
- The *k* random samples are ... **independent**
- For each of population the model for the response is... **a normal distribution**
- The *k* population variances are .... **equal**.

State the hypotheses to be tested:

H_0: ______________________ versus H_a:__________________________________

Note: We would use a computer or calculator to work at least the basic summaries in steps 1 and 2, and likely to create the entire ANOVA table for us. Let's be sure we understand where the values are coming from and how to interpret the final results.

Step 1: Calculate the mean and variance for each sample:

$\bar{x}_1 =$ $\qquad\qquad$ $s_1^2 =$

$\bar{x}_2 =$ $\qquad\qquad$ $s_2^2 =$

$\bar{x}_3 =$ $\qquad\qquad$ $s_3^2 =$

Step 2: Calculate the overall sample mean
(based on all $N = n_1 + n_2 + \ldots + n_k$ observations):

$\bar{x} =$

Step 3: Calculate the sum of squares between groups:

$$\text{SS Groups} = \sum_{groups} n_i (\bar{x}_i - \bar{x})^2$$

Step 4: Calculate the sum of squares within groups (due to error):

$$\text{SSE} = \sum_{groups} (n_i - 1) s_i^2$$

Step 5: OPTIONAL: Calculate the total sum of squares: *No Thank You*!

Step 6: Fill in the ANOVA table:

Source	Sum of Squares	DF	Mean Square	F
Groups				
Error (Within)				
Total				

Here are the results from SPSS:

ANOVA

RESPONSE

	Sum of Squares	df	Mean Square	F	Sig.
Between Groups	21.981	2	10.991	4.188	.034
Within Groups	41.988	16	2.624		
Total	63.969	18			

One of the assumptions in ANOVA is that the population standard deviations are all equal. Using the data, give an estimate of this common population standard deviation.

Give the observed test statistic value.

What is the distribution of the test statistic if the three drugs are equally effective in terms of the mean response?

What is the corresponding *p*-value for assessing if the three drugs are equally effective in terms of the mean response?

At the 5% level, what is your conclusion?

We Rejected H_0 in ANOVA: What is next?
Multiple Comparisons

The term **multiple comparisons** is used when two or more comparisons are made to examine the specific pattern of differences among means. The most commonly analyzed set of multiple comparisons is the set of all pairwise comparisons among population means. In our previous Drug example, the possible pairwise comparisons are: Drug 1 with Drug 2, Drug 1 with Drug 3, and Drug 2 with Drug 3.

To compare the pair of means we could ...

- Compute a **confidence interval** for the difference between the two population means and see if 0 falls in the interval or not.

- Perform a **test of hypotheses** to assess if the two population means differ significantly.

When many statistical tests are done there is an increased risk of making at least one type I error (erroneously rejecting a null hypothesis). Consequently, several procedures have been developed to **control the overall family type I error rate** or the **overall family confidence level** when inferences for a set (family) of multiple comparisons are done.

Tukey's procedure is one such procedure for the family of pairwise comparisons. If the family error rate is not a concern, Fisher's procedure is used.

Try It! Comparing 3 Drugs

In the comparison of the three drugs, we rejected the null hypothesis at the 5% significance level. We follow with a multiple comparison procedure to determine which group means are significantly different from each other.

SPSS multiple comparisons results using Tukey's method and a family confidence level of 95% are provided on the next page.

Multiple Comparisons

Dependent Variable: TIME
Tukey HSD

(I) DRUG	(J) DRUG	Mean Difference (I-J)	Std. Error	Sig.	95% Confidence Interval	
					Lower Bound	Upper Bound
1	2	-1.0800	.9485	.505	-3.5276	1.3676
	3	1.4200	.9485	.318	-1.0276	3.8676
2	1	1.0800	.9485	.505	-1.3676	3.5276
	3	2.5000*	.8659	.027	.2657	4.7343
3	1	-1.4200	.9485	.318	-3.8676	1.0276
	2	-2.5000*	.8659	.027	-4.7343	-.2657

*. The mean difference is significant at the .05 level.

a. Use the above output to report about the three pairwise comparisons:

Does the confidence interval for $\mu_1 - \mu_2$ contain 0? __________

Does the confidence interval for $\mu_1 - \mu_3$ contain 0? __________

Does the confidence interval for $\mu_2 - \mu_3$ contain 0? __________

b. State your conclusions regarding the differences between the mean response for the three drug groups based on the Tukey multiple comparison method.

We can conclude that the population mean responses differ for ...

but do not differ for ...

Individual Confidence Intervals for the Population Means

Sometimes it is helpful to examine a confidence interval for the mean for each population. Since in ANOVA we assume the population standard deviations are all equal, the estimate of that common population standard deviation $s_p = \sqrt{MSE}$ is used in forming the individual confidence intervals. The degrees of freedom used to find the t^* multiplier will be those associated with the estimated standard deviation, namely $N - k$. The summary for the individual confidence intervals is provided below.

In one-way analysis of variance, a **confidence interval for a population mean** μ_i is

$$\bar{x}_i \pm t^* \frac{s_p}{\sqrt{n_i}}$$

where $s_p = \sqrt{\text{MSE}}$ and t^* is such that the confidence level is the probability between $-t^*$ and t^* in a t-distribution with $N - k$ degrees of freedom.

From Utts, Jessica M. and Robert F. Heckard. Mind on Statistics, Fourth Edition. 2012. Used with permission.

Try It! Comparing 3 Drugs

We were comparing k = 3 groups based on a total of N = 19 observations. The pooled standard deviation for the comparison of the three drugs data set is s_p = 1.62. The sample means and sample sizes were:

Drug 1:	Sample mean = 8.22	Sample size = 5
Drug 2:	Sample mean = 9.30	Sample size = 7
Drug 3:	Sample mean = 6.80	Sample size = 7

The degrees of freedom for the t^* multiplier is $N - k$ = ____________________.

From the table of t^* multipliers (Table A.2) with confidence level = 0.95 and the above degrees of freedom we have t^* = ________________________

Drug 3 was descriptively the best. Compute a 95% confidence interval for the population mean response for all subjects taking Drug 3.

Try It! Memory Experiment

In a memory experiment, three groups of subjects were given a list of words to try to remember. The length of the list for the first group was 10 words (short list), whereas for the second group it was 20 words (medium list) and for the third group 40 words (long list). The percentage of words recalled for each subject was recorded. The sample mean percentage of words recalled was 68.3% for the short list, 48% for the medium list, and 39.2%for the long list. A one-way ANOVA was used to assess whether the *length of the list* had a significant effect on the *percentage of words recalled.*

	SS	df	Mean Square	F	Sig.
Between Groups	2668.8	2			.0003
Within Groups			84.6		
Total	3852.9	16			

a. Some values in the ANOVA table are missing. Complete the above table.

b. State the null and alternative hypotheses that the above F statistic is testing.

H_0: ____________________ vs. H_a: ____

c. Suppose the necessary assumptions hold. Using a 5% significance level, does it appear that the average percentage of words recalled is the same for the three different lengths of lists? Explain.

d. Multiple comparisons were performed using Tukey's method.

Use the results and circle the pairs that are significantly different at a 1% level.

(I) GROUP	(J) GROUP	Mean Difference (I-J)	99% Confidence Interval Lower Bound	99% Confidence Interval Upper Bound
short list	medium list	20.33	1.06	39.61
	long list	29.17	10.79	47.54
medium list	short list	-20.33	-39.61	-1.06
	long list	8.83	-10.44	28.11
long list	short list	-29.17	-47.54	-10.79
	medium list	-8.83	-28.11	10.44

short versus medium **short versus long** **medium versus long**

e. Give a 99% confidence interval for the population mean percentage of words recalled for the long list group. Recall that the sample mean based on the 6 subjects in the long list group was 39.2 percent.

What if some conditions do not hold?

You probably won't be surprised to learn that the necessary conditions for using an analysis of variance *F*-test don't hold for all data sets. In Section 16.3, two methods are discussed that can be used when one or both of the assumptions about equal population standard deviations and normal distributions are violated.

When the observed **data are skewed**, or when extreme outliers are present, it usually is better to **analyze the median rather than the mean**. One test for comparing medians is the **Kruskal-Wallis Test**. It is based on a comparison of the relative rankings (sizes) of the data in the observed samples, and for this reason is called a rank test. The term **nonparametric test** also is used to describe this test because there are no assumptions made about a specific distribution for the population of measurements. Another nonparametric test used to compare population medians is **Mood's Median Test**. See Section 16.3 for an example of each of these alternative methods.

Two-Way ANOVA

So far we have focused on the **one-way ANOVA** procedure. The "*one-way*" referred to having only one explanatory variable (or factor) and one quantitative response variable.

Section 16.4 presents a brief overview of **two-way ANOVA**, which examines the effect of two explanatory variables (or factors) on the mean response. The researcher is interested in the *individual effect* of each explanatory variable on the mean response and also in the *combined effect* of the two explanatory variables on the mean response. The individual effect of each factor on the response is called a **main effect**. If one of the factors does not have an effect on the response, we say there is no main effect due to that factor.

Besides assessing the main effects of each factor on the response, an interesting feature in two-way analyses is the possibility of interaction between the two factors. We say **there is interaction between two factors if the effect of one factor on the mean response depends on the specific level of the other factor**. The interpretation of the factor main effects can be more difficult when interaction is present.

Two-Way Anova will not be discussed in our class. The next course to follow Stat 250, which would include the modeling for interaction, is Statistics 401 (generally offered both Fall and Winter terms).

Additional Notes

A place to … jot down questions you may have and ask during office hours, take a few extra notes, write out an extra practice problem or summary completed in lecture, create your own short summary about this chapter.

Stat 250 Gunderson Lecture Notes
Chapters 3 and 14: Regression Analysis

> The invalid assumption that correlation implies cause is probably among the two or three most serious and common errors of human reasoning.
>
> --*Stephen Jay Gould, The Mismeasure of Man* © FAIR USE

Describing and assessing the significance of **relationships between variables** is very important in research. We will first learn how to do this in the case when the two variables are quantitative. Quantitative variables have numerical values that can be ordered according to those values. We will study the material from Chapters 3 and 14 together. We will merge the two chapters together into one overall discussion of these ideas.

Main idea

We wish to study the relationship between two quantitative variables.

Generally one variable is the _________________________ ***variable***, denoted by *y*. This variable measures the outcome of the study and is also called the _________________________ variable.

The other variable is the _________________________ ***variable***, denoted by *x*. It is the variable that is thought to explain the changes we see in the response variable. The explanatory variable is also called the _________________ variable.

The first step in examining the relationship is to use a graph - a ***scatterplot*** - to display the relationship. We will look for an overall pattern and see if there are any departures from this overall pattern.

If a ***linear*** relationship appears to be reasonable from the scatterplot, we will take the next step of finding a model (an equation of a line) to summarize the relationship. The resulting equation may be used for predicting the response for various values of the explanatory variable. If certain assumptions hold, we can assess the significance of the linear relationship and make some confidence intervals for our estimations and predictions.

Let's begin with an example that we will carry throughout our discussions.

Graphing the Relationship: Exam 2 versus Final Scores

How well does the exam 2 score for a Stats 250 student predict their final exam score? Below are the scores for a random sample of *n* = 6 students from a previous term.

Exam 2 Score	33	65	44	64	60	40
Final Exam Score	53	80	78	93	88	58

Response (dependent) variable y = ______________________________ .

Explanatory (independent) variable x = ______________________________ .

Step 1: Examine the data graphically with a scatterplot.

Add the points to the scatterplot below:

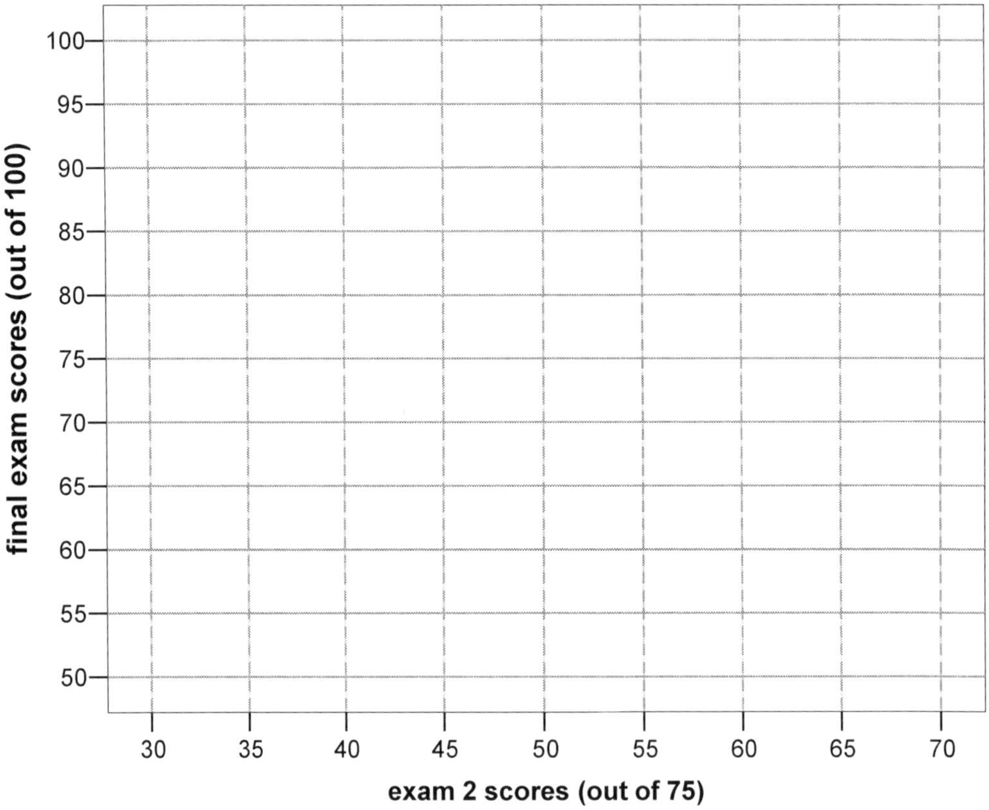

Interpret the scatterplot in terms of ...

- **overall form**
  (is the average pattern look like a straight line or is it curved?)
- **direction** of association (positive or negative)
- **strength** of association
  (how much do the points vary around the average pattern?)
- any **deviations** from the overall form?

Describing a Linear Relationship with a Regression Line

Regression analysis is the area of statistics used to examine the relationship between a quantitative response variable and one or more explanatory variables. A key element is the **estimation of an equation** that describes how, on average, the response variable is related to the explanatory variables. A regression equation can also be used to make predictions.

The simplest kind of relationship between two variables is a straight line, the analysis in this case is called **linear regression.**

Regression Line for Exam 2 vs Final

Remember the equation of a line?
In statistics we denote the **regression line for a sample** as:
where:

$\hat{y}$

b_0

b_1

Goal:
To find a line that is "close" to the data points - find the "best fitting" line.

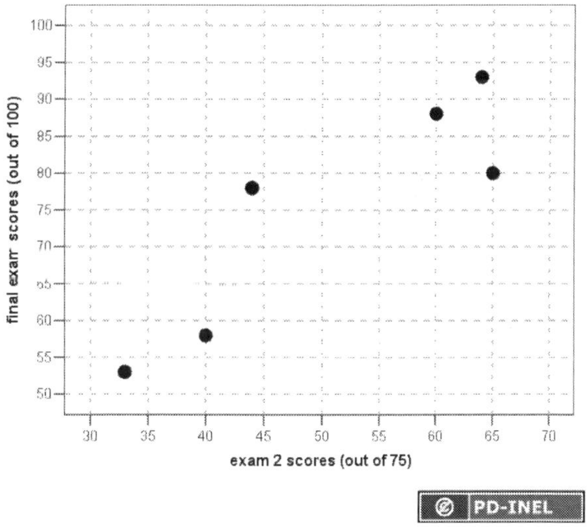

How?
What do we mean by best?

One measure of how good a line fits is to look at the "observed errors" in prediction.

Observed errors = ______________________ are called ____________________

So we want to choose the line for which the sum of squares of the observed errors (the sum of squared residuals) is the ***least***.

The line that does this is called:____

The equations for the estimated slope and intercept are given by:

$b_1 =$

$b_0 =$

The least squares regression line (estimated regression function) is:
$\hat{y} = \hat{\mu}_y(x) = b_0 + b_1 x$

To find this estimated regression line for our exam data by hand, it is easier if we set up a calculation table. By filling in this table and computing the column totals, we will have all of the main summaries needed to perform a complete linear regression analysis. Note that here we have $n = 6$ observations. The first five rows have been completed for you. In general, use SPSS or a calculator to help with the graphing and numerical computations!

x = exam 2	y = final	$x - \bar{x}$	$(x - \bar{x})^2$	$(x - \bar{x})y$	$y - \bar{y}$	$(y - \bar{y})^2$
33	53	33–51 = -18	$(-18)^2 = 324$	(-18)(53)= - 954	53–75 = -22	$(-22)^2 = 484$
65	80	65–51= 14	$(14)^2 = 196$	(14)(80)= 1120	80–75 = 5	$(5)^2 = 25$
44	78	44–51= -7	$(-7)^2 = 49$	(-7)(78)= -546	78–75 = 3	$(3)^2 = 9$
64	93	64–51= 13	$(13)^2 = 169$	(13)(93)= 1209	93–75 = 18	$(18)^2 = 324$
60	88	60–51 = 9	$(9)^2 = 81$	(9)(88)= 792	88–75 = 13	$(13)^2 = 169$
40	58					
306	**450**					

$\bar{x} = \frac{306}{6} = 51 \quad \bar{y} = \frac{450}{6} = 75$

Slope Estimate:

y-intercept Estimate:

Estimated Regression Line:

Predict the final exam score for a student who scored 60 on exam 2.

Note: The 5th student had an exam 2 score of 60 and the observed final exam score was 88 points.

Find the residual for the 5th observation.

Notation for a residual $= e_5 = y_5 - \hat{y}_5 =$

The residuals ...

You found the residual for one observation. You could compute the residual for each observation. The following table shows each residual.

x = exam 2	y = final	predicted values $\hat{y} = 21.67 + 1.046(x)$	residuals $e = y - \hat{y}$	Squared residuals $(e)^2 = (y - \hat{y})^2$
33	53	56.19	-3.19	10.18
65	80	89.66	-9.66	93.31
44	78	67.69	10.31	106.29
64	93	88.61	4.39	19.27
60	88	84.43	3.57	12.74
40	58	63.51	-5.51	30.36
--	--	--		

SSE = sum of squared errors (or residuals) ≈

Measuring Strength and Direction of a Linear Relationship with Correlation

The ***correlation coefficient r*** is a measure of strength of the linear relationship between y and x.

Properties about the Correlation Coefficient *r*

1. r ranges from ...

2. Sign of r indicates ...

3. Magnitude of r indicates ...

   A "strong" r is discipline specific
   - $r = 0.8$ might be an important (or strong) correlation in engineering
   - $r = 0.6$ might be a strong correlation in psychology or medical research

4. r ONLY measures the strength of the ____________________ relationship.

Some pictures:

The **formula** for the correlation:
(but we will get it from computer output or from r^2)

$$r = \frac{1}{n-1}\sum_i \left(\frac{x_i - \bar{x}}{s_x}\right)\left(\frac{y_i - \bar{y}}{s_y}\right)$$

Exam Scores Example:

$r =$ ____________

Interpretation:

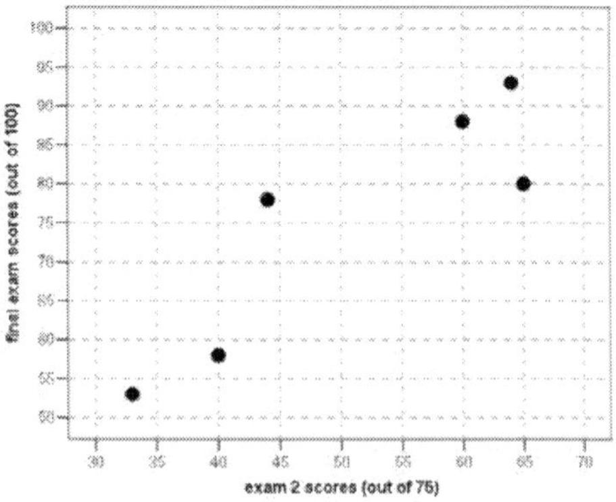

The square of the correlation r^2

The squared correlation coefficient r^2 always has a value between __________ and is sometimes presented as a percent. It can be shown that the square of the correlation is related to the sums of squares that arise in regression.

The responses (the final exam scores) in data set are not all the same - they do vary. We would measure the ***total variation*** in these responses as $\text{SSTO} = \sum(y - \bar{y})^2$ (last column total in calculation table said we would use later).

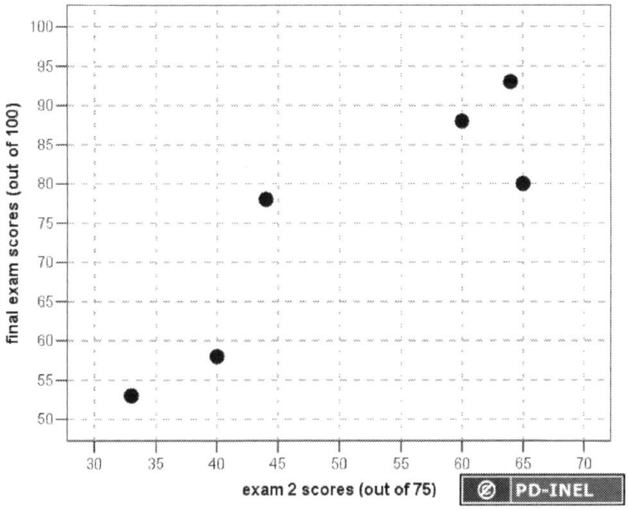

Part of the reason why the final exam scores vary is because there is a linear relationship between final exam scores and exam 2 scores, and the study included students with different exam 2 scores. When we found the least squares regression line, there was still some small variation remaining of the responses from the line. This amount of **variation that is not accounted for by the linear relationship** is called the **SSE**.

The amount of **variation that is accounted for by the linear relationship** is called the sum of squares due to the model (or regression), denoted by **SSM** (or sometimes as SSR).

So we have: **SSTO = ____________________________________**

It can be shown that

r^2 =

= the proportion of total variability in the responses that can be explained by the linear relationship with the explanatory variable x.

Note: The value of r^2 and these sums of squares are summarized in an ***ANOVA table*** that is standard output from computer packages when doing regression.

Measuring Strength and Direction for Exam 2 vs Final

From our first calculation table we have:

SSTO = ______________________________________

From our residual calculation table we have:

SSE = _____________________________________

So the squared correlation coefficient for our exam scores regression is:

$$r^2 = \frac{SSTO - SSE}{SSTO} =$$

Interpretation:

We accounted for _____________ % of the variation in

by the linear regression on __________________ .

The correlation coefficient is r = ___________________

Be sure you read Sections 3.4 and 3.5 (pages 89 – 95 of the textbook) for good examples and discussion on the following topics:

- Nonlinear relationships
- Detecting Outliers and their influence on regression results.
- Dangers of Extrapolation (predicting outside the range of your data)
- Dangers of combining groups inappropriately (Simpson's Paradox)
- Correlation does not prove causation

SPSS Regression Analysis for Exam 2 vs Final

Let's look at the SPSS output for our exam data.
We will see that much of the computations are done for us.

Variables Entered/Removed[b]

Model	Variables Entered	Variables Removed	Method
1	exam 2 scores (out of 75)[a]	.	Enter

a. All requested variables entered.

b. Dependent Variable: final exam scores (out of 100)

Model Summary[b]

Model	R	R Square	Adjusted R Square	Std. Error of the Estimate
1	.889[a]	.791	.738	8.24671

a. Predictors: (Constant), exam 2 scores (out of 75)

b. Dependent Variable: final exam scores (out of 100)

ANOVA[b]

Model		Sum of Squares	df	Mean Square	F	Sig.
1	Regression	1027.967	1	1027.967	15.115	.018[a]
	Residual	272.033	4	68.008		
	Total	1300.000	5			

a. Predictors: (Constant), exam 2 scores (out of 75)

b. Dependent Variable: final exam scores (out of 100)

Coefficients[a]

Model		Unstandardized Coefficients		Standardized Coefficients	t	Sig.
		B	Std. Error	Beta		
1	(Constant)	21.667	14.125		1.534	.200
	exam 2 scores (out of 75)	1.046	.269	.889	3.888	.018

a. Dependent Variable: final exam scores (out of 100)

Inference in Linear Regression Analysis

The material covered so far is presented in Chapter 3 and focuses on using the data for a **sample** to graph and describe the relationship. The slope and intercept values we have computed are statistics, they are estimates of the underlying true relationship for the larger population.

Chapter 14 focuses on making inferences about the relationship for the larger **population**. Here is a nice summary to help us distinguish between the **regression line for the sample** and the **regression line for the population**.

Regression Line for the Sample

$$\hat{y} = b_0 + b_1 x$$

In any given situation, the sample is used to determine values for b_0 and b_1.

- $\hat{y}$ is spoken as "y-hat" and it is also referred to either as *predicted y* or *estimated y*.
- b_0 is the **intercept** of the straight line. The *intercept* is the value of $\hat{y}$ when $x = 0$.
- b_1 is the **slope** of the straight line. The *slope* tells us how much of an increase (or decrease) there is for $\hat{y}$ when the x variable increases by one unit. The sign of the slope tells us whether $\hat{y}$ increases or decreases when x increases. If the slope is 0, there is no linear relationship between x and y because $\hat{y}$ is the same for all values of x.

Regression Line for the Population

The regression equation for a simple linear relationship in a population can be written as

$$E(Y) = \beta_0 + \beta_1 x$$

- $E(Y)$ represents the mean or expected value of y for individuals in the population who all have the same particular value of x. Note that $\hat{y}$ is an estimate of $E(Y)$.
- β_0 is the **intercept** of the straight line in the **population.**
- β_1 is the **slope** of the line in the **population.** Note that if the slope $\beta_1 = 0$, there is no linear relationship in the population.

To do formal inference, we think of our b_0 and b_1 as estimates of the unknown parameters β_0 and β_1 . Below we have the somewhat statistical way of expressing the underlying model that produces our data:

Linear Model: the response $y = [\beta_0 + \beta_1(x)] + \varepsilon$
$= [\text{Population relationship}] + \text{Randomness}$

This statistical model for simple linear regression assumes that for each value of x the observed values of the response (the population of y values) is **normally distributed**, varying around some true **mean (that may depend on x in a linear way)** and a **standard deviation σ that does not depend on x**. This true mean is sometimes expressed as $E(Y) = \beta_0 + \beta_1(x)$. And the components and assumptions regarding this statistical model are show visually below.

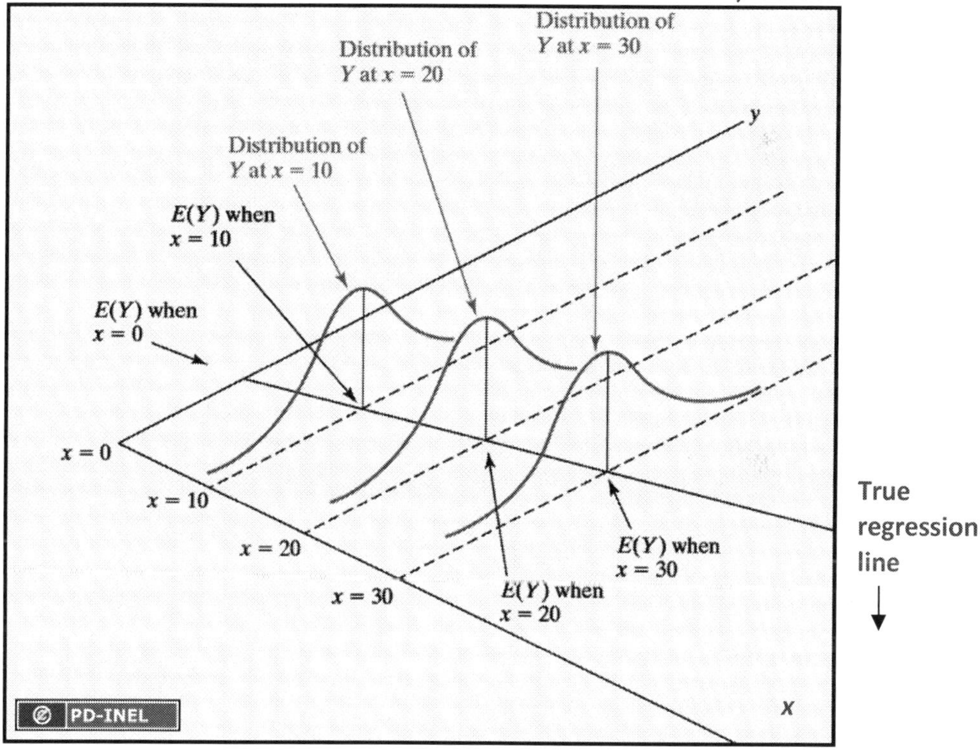

The ε represents the *true error term.* These would be the deviations of a particular value of the response y from the *true regression line.* As these are the deviations from the mean, then these error terms should have a normal distribution with mean 0 and constant standard deviation σ.

Now, we cannot observe these ε's. However we will be able to use the estimated (observable) errors, namely the residuals, to come up with an estimate of the standard deviation and to check the conditions about the true errors.

So what have we done, and where are we going?

1. Estimate the regression line based on some data. **DONE!**
2. Measure the strength of the linear relationship with the correlation. **DONE!**
3. Use the estimated equation for predictions. **DONE!**
4. Assess if the linear relationship is statistically significant.
5. Provide interval estimates (confidence intervals) for our predictions.
6. Understand and check the assumptions of our model.

We have already discussed the descriptive goals of 1, 2, and 3. For the inferential goals of 4 and 5, we will need an estimate of the unknown standard deviation in regression σ.

Estimating the Standard Deviation for Regression

The standard deviation for regression can be thought of as measuring the **average size of the residuals**. A relatively small standard deviation from the regression line indicates that individual data points generally fall close to the line, so predictions based on the line will be close to the actual values.

It seems reasonable that our estimate of this average size of the residuals be based on the residuals using the sum of squared residuals and dividing by appropriate degrees of freedom. Our estimate of σ is given by:

$s =$

Note: Why $n - 2$?

Estimating the Standard Deviation: Exam 2 vs Final

Below are the portions of the SPSS regression output that we could use to obtain the estimate of σ for our regression analysis.

Model Summary [b]

Model	R	R Square	Adjusted R Square	Std. Error of the Estimate
1	.889 [a]	.791	.738	8.24671

a. Predictors: (Constant), exam2 scores (out of 75)

b. Dependent Variable: final exam scores (out of 100)

ANOVA [b]

Model		Sum of Squares	df	Mean Square	F	Sig.
1	Regression	1027.967	1	1027.967	15.115	.018 [a]
	Residual	272.033	4	68.008		
	Total	1300.000	5			

a. Predictors: (Constant), exam2 scores (out of 75)

b. Dependent Variable: final exam scores (out of 100)

Significant Linear Relationship?

Consider the following hypotheses: $H_0 : \beta_1 = 0$ versus $H_a : \beta_1 \neq 0$

What happens if the null hypothesis is true?

There are a number of ways to test this hypothesis. One way is through a t-test statistic (think about why it is a t and not a z test).

The general form for a t test statistic is:

$$\text{t} = \frac{\text{sample statistic - null value}}{\text{standard error of the sample statistic}}$$

We have our sample estimate for β_1, it is b_1. And we have the null value of 0. So we need the standard error for b_1. We could "derive" it, using the idea of sampling distributions (think about the population of all possible b_1 values if we were to repeat this procedure over and over many times). Here is the result:

t-test for the population slope β_1

To test $H_0 : \beta_1 = 0$ we would use $t = \dfrac{b_1 - 0}{\text{s.e.}(b_1)}$

where $SE(b_1) = \dfrac{s}{\sqrt{\sum(x - \bar{x})^2}}$

and the degrees of freedom for the *t*-distribution are $n - 2$.

This t-statistic could be modified to test a variety of hypotheses about the population slope (different null values and various directions of extreme).

Try It!

Significant Relationship between Exam 2 Scores and Final Scores?

Is there a significant (non-zero) linear relationship between the exam 2 scores and the final exam scores? (is exam 2 a useful linear predictor for final exam score?)

That is, test $H_0 : \beta_1 = 0$ versus $H_a : \beta_1 \neq 0$ using a 5% level of significance.

Think about it:

Based on the results of the previous *t*-test conducted at the 5% significance level, do you think a 95% confidence interval for the true slope β_1 would contain the value of 0?

Confidence Interval for the population slope β_1

$b_1 \pm t^* [SE(b_1)]$ where df = $n - 2$ for the t^* value

Compute the interval and check your answer.

Could you interpret the 95% confidence level here?

Inference about the Population Slope using SPSS

Below are the portions of the SPSS regression output that we could use to perform the *t*-test and obtain the confidence interval for the population slope β_1.

Coefficients[a]

Model		Unstandardized Coefficients B	Unstandardized Coefficients Std. Error	Standardized Coefficients Beta	t	Sig.
1	(Constant)	21.667	14.125		1.534	.200
	exam 2 scores (out of 75)	1.046	.269	.889	3.888	.018

a. Dependent Variable: final exam scores (out of 100)

Note: There is a third way to test $H_0 : \beta_1 = 0$ versus $H_a : \beta_1 \neq 0$.

It involves another *F*-test from an ANOVA for regression.

ANOVA[b]

Model		Sum of Squares	df	Mean Square	F	Sig.
1	Regression	1027.967	1	1027.967	15.115	.018[a]
	Residual	272.033	4	68.008		
	Total	1300.000	5			

a. Predictors: (Constant), exam 2 scores (out of 75)

b. Dependent Variable: final exam scores (out of 100)

Predicting for Individuals versus Estimating the Mean

Consider the relationship between exam 2 and final exam scores ...

Least squares regression line (or estimated regression function):

$$\hat{y} =$$

We also have: $s =$

How would you predict the final exam score ***for Barb***
who scored 60 points on exam 2?

How would you estimate the mean final exam score ***for all students***
who scored 60 points on exam 2?

So our estimate for **predicting a future observation** and for **estimating the mean response** are found using the same least squares regression equation. What about their standard errors? (We would need the standard errors to be able to produce an interval estimate.)

Idea: Consider a population of individuals and a population of means:

What is the standard deviation for a population of individuals?

What is the standard deviation for a population of means?
Which standard deviation is larger?

So a ***prediction interval for an individual response*** will be

(wider or narrower) than a ***confidence interval for a mean response***.

Here are the (somewhat messy) formulas:

Confidence interval for a mean response: $\hat{y} \pm t^{*}\text{s.e.(fit)}$

where $\text{s.e.(fit)} = s\sqrt{\frac{1}{n} + \frac{(x-\bar{x})^2}{\sum(x_i - \bar{x})^2}}$ **df = *n* – 2**

Prediction interval for an individual response: $\hat{y} \pm t^{*}\text{s.e.(pred)}$

where $\text{s.e.(pred)} = \sqrt{s^2 + (\text{s.e.(fit)})^2}$ **df = *n* – 2**

Try It! Exam 2 vs Final

Construct a 95% confidence interval for the mean final exam score for all students who scored *x* = 60 points on exam 2. Recall: $n = 6$, $\bar{x} = 51$, $\sum(x-\bar{x})^2 = S_{XX} = 940$, $\hat{y} = 21.67 + 1.046(x)$, and $s = 8.24761$.

Construct a 95% prediction interval for the final exam score for an individual student who scored *x* = 60 points on exam 2.

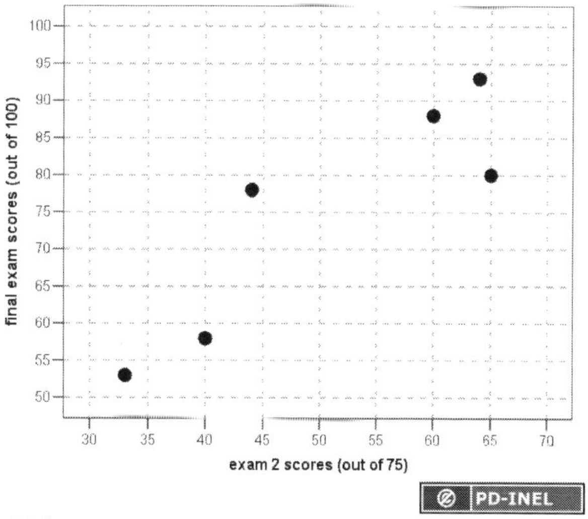

Checking Assumptions in Regression

Let's recall the statistical way of expressing the underlying model that produces our data:

Linear Model: the response $y = [\beta_0 + \beta_1(x)] + \varepsilon$
= [Population relationship] + Randomness

where the ε's, the *true error terms* should be normally distributed with mean 0 and constant standard deviation σ, and this randomness is independent from one case to another.

Thus there are **four essential technical assumptions** required for inference in linear regression:

(1) Relationship is in fact linear.
(2) Errors should be normally distributed.
(3) Errors should have constant variance.
(4) Errors should not display obvious 'patterns'.

Now, we cannot observe these ε's. However we will be able to use the estimated (observable) errors, namely the residuals, to come up with an estimate of the standard deviation and to check the conditions about the true errors.

So how can we check these assumptions with our data and estimated model?

(1) Relationship is in fact linear. →

(2) Errors should be normally distributed. →

(3) Errors should have constant variance.
(4) Errors should not display obvious 'patterns'. }

If we saw ...

Let's turn to one last full regression problem that includes checking assumptions.

Relationship between height and foot length for College Men

The heights (in inches) and foot lengths (in centimeters) of 32 college men were used to develop a model for the relationship between height and foot length. The scatterplot and SPSS regression output are provided.

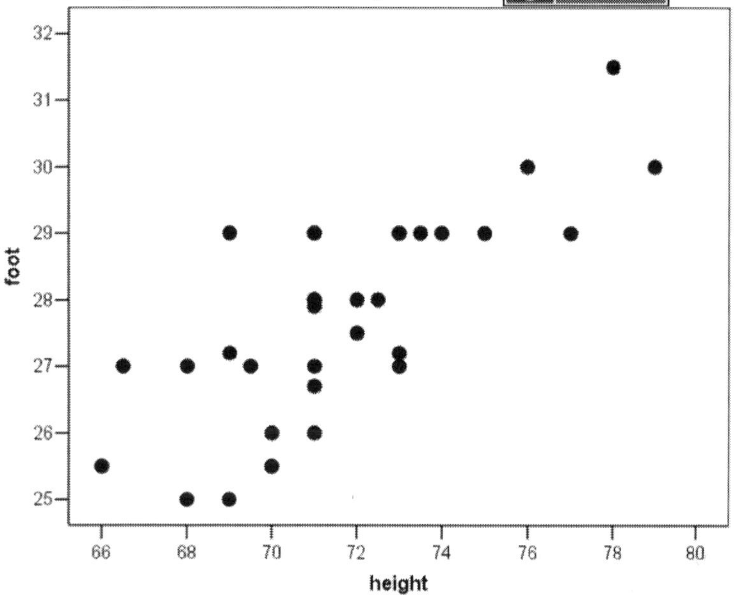

Descriptive Statistics

	Mean	Std. Deviation	N
foot	27.8	1.5497	32
height	71.7	3.0579	32

Model Summary

Model	R	R Square	Adjusted R Square	Std. Error of the Estimate
1	.758	.574	.560	1.0280

ANOVA

Model		Sum of Squares	df	Mean Square	F	Sig.
1	Regression	42.74	1	42.74	40.45	.000001
	Residual	31.70	30	1.06		
	Total	74.45	31			

Coefficients

Model		Unstandardized Coefficients		Standardized Coefficients	t	Sig.
		B	Std. Error	Beta		
1	(Constant)	.25	4.33		.06	.954
	height	.38	.06	.758	6.36	.000001

Also note that: $S_{XX} = \sum(x - \bar{x})^2 = 289.87$

a. How much would you expect foot length to increase for each 1-inch increase in height? Include the units.

b. What is the correlation between height and foot length?

c. Give the equation of the least squares regression line for predicting foot length from height.

d. Suppose Max is 70 inches tall and has a foot length of 28.5 centimeters. Based on the least squares regression line, what is the value of the prediction error (residual) for Max? Show all work.

e. Use a 1% significance level to assess if there is a significant positive linear relationship between height and foot length. State the hypotheses to be tested, the observed value of the test statistic, the corresponding *p*-value, and your decision.

Hypotheses: H_0:______________________

H_a:______________________

Test Statistic Value: ________________ *p*-value: ____________________

Decision: (circle) **Fail to reject H_0** **Reject H_0**

Conclusion:

f. Calculate a 95% confidence interval for the average foot length for all college men who are 70 inches tall. (Just clearly plug in all numerical values.)

g. Consider the residual plot shown.

PD-INEL

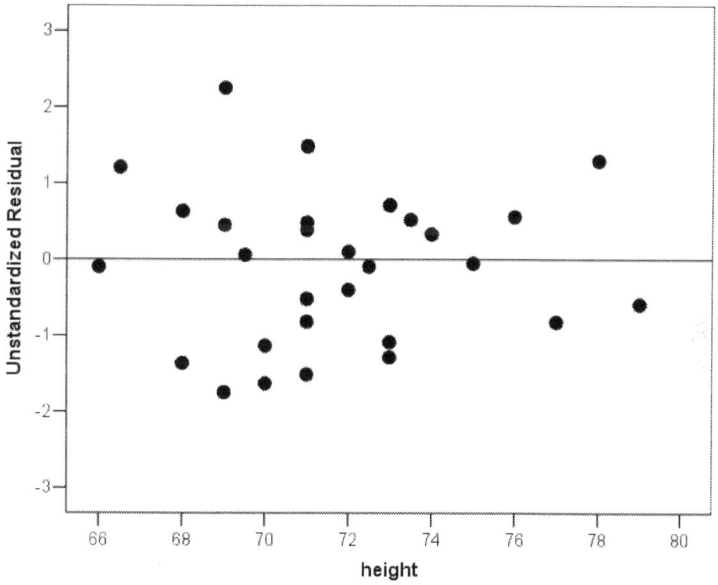

Does this plot support the conclusion that the linear regression model is appropriate? ***Yes*** ***No***

Explain:

Regression

Linear Regression Model

Population Version:

Mean: $\mu_Y(x) = E(Y) = \beta_0 + \beta_1 x$

Individual: $y_i = \beta_0 + \beta_1 x_i + \varepsilon_i$

where ε_i is $N(0, \sigma)$

Sample Version:

Mean: $\hat{y} = b_0 + b_1 x$

Individual: $y_i = b_0 + b_1 x_i + e_i$

Standard Error of the Sample Slope

$$\text{s.e.}(b_1) = \frac{s}{\sqrt{S_{XX}}} = \frac{s}{\sqrt{\sum (x - \bar{x})^2}}$$

Confidence Interval for β_1

$b_1 \pm t^* \text{s.e.}(b_1)$ df = $n - 2$

***t*-Test for β_1** To test $H_0 : \beta_1 = 0$

$$t = \frac{b_1 - 0}{\text{s.e.}(b_1)}$$ df = $n - 2$

or $F = \dfrac{MSREG}{MSE}$ df = $1, n - 2$

Parameter Estimators

$$b_1 = \frac{S_{XY}}{S_{XX}} = \frac{\sum (x - \bar{x})(y - \bar{y})}{\sum (x - \bar{x})^2} = \frac{\sum (x - \bar{x}) y}{\sum (x - \bar{x})^2}$$

$$b_0 = \bar{y} - b_1 \bar{x}$$

Confidence Interval for the Mean Response

$\hat{y} \pm t^* \text{s.e.(fit)}$ df = $n - 2$

where $\text{s.e.(fit)} = s\sqrt{\dfrac{1}{n} + \dfrac{(x - \bar{x})^2}{S_{XX}}}$

Residuals

$e = y - \hat{y}$ = observed y – predicted y

Prediction Interval for an Individual Response

$\hat{y} \pm t^* \text{s.e.(pred)}$ df = $n - 2$

where $\text{s.e.(pred)} = \sqrt{s^2 + (\text{s.e.(fit)})^2}$

Correlation and its square

$$r = \frac{S_{XY}}{\sqrt{S_{XX} S_{YY}}}$$

$$r^2 = \frac{SSTO - SSE}{SSTO} = \frac{SSREG}{SSTO}$$

where $SSTO = S_{YY} = \sum (y - \bar{y})^2$

Standard Error of the Sample Intercept

$$\text{s.e.}(b_0) = s\sqrt{\frac{1}{n} + \frac{\bar{x}^2}{S_{XX}}}$$

Confidence Interval for β_0

$b_0 \pm t^* \text{s.e.}(b_0)$ df = $n - 2$

Estimate of σ

$s = \sqrt{MSE} = \sqrt{\dfrac{SSE}{n-2}}$ where

$$SSE = \sum (y - \hat{y})^2 = \sum e^2$$

***t*-Test for β_0** To test $H_0 : \beta_0 = 0$

$$t = \frac{b_0 - 0}{\text{s.e.}(b_0)}$$ df = $n - 2$

Additional Notes

A place to ... jot down questions you may have and ask during office hours, take a few extra notes, write out an extra practice problem or summary completed in lecture, create your own short summary about this chapter.

Stat 250 Gunderson Lecture Notes
Relationships between Categorical Variables
Chapter 15: Chi-Square Analysis

Inference for Categorical Variables

Chapter 15 teaches us the inference techniques for analyzing count data. The three main tests described in the text that we will cover are:

1. ***Goodness of Fit Test:*** this test is for assessing if a particular discrete model is a good fitting model for a discrete characteristic, based on a random sample from the population.
   E.g. Has the model for the method of transportation (drive, bike, walk, other) used by students to get the class changed from that for 5 years ago?

2. ***Test of Homogeneity***: this test is for assessing if two or more populations are homogeneous (alike) with respect to the distribution of some discrete (categorical) variable.
   E.g. Is the distribution of opinion on legal gambling the same for adult males versus adult females?

3. ***Test of Independence***: this test helps us to assess if two discrete (categorical) variables are independent for a population, or if there is an association between the two variables.
   E.g. Is there an association between satisfaction with the quality of public schools (not satisfied, somewhat satisfied, very satisfied) and political party (Republican, Democrat, etc.)

The first test is discussed in Section 15.3 in the text but we will cover it first as it is the one-sample test for count data.

The other two tests (homogeneity and independence) are actually the same test. Although the hypotheses are stated differently and the underlying assumptions about how the data is gathered are different, the steps for doing the two tests are exactly the same. Section 15.1 does not fully distinguish between these two tests (just briefly on page 586 of the textbook) but we will emphasize the difference between them.

All of these tests are based on an X^2 test statistic that, if the corresponding H_0 is true and the assumptions hold, follows a **chi-square distribution** with some degrees of freedom, written $\chi^2(df)$.

So our first discussion is to learn about the chi-square distribution - what the distribution looks like, some facts, how to use Table A.5 to find various percentiles.

The Chi-Square Distribution

General Shape:
If we have a chi-square distribution with df = degrees of freedom, then the ...

Mean is equal to __________

Variance is equal to __________

Standard deviation is equal to ________

These facts will serve as a useful frame of reference for making decision.

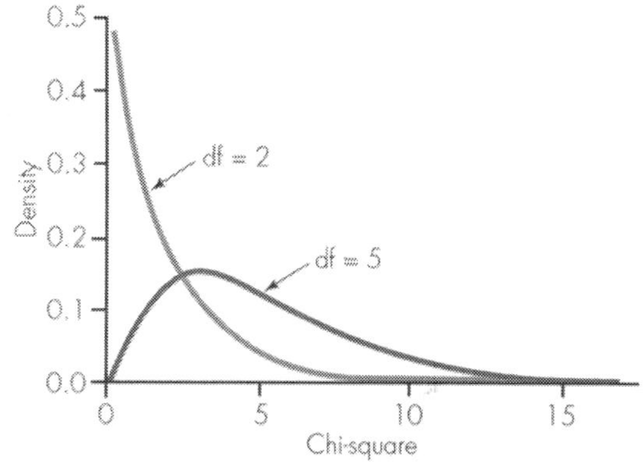

Figure 15.2 Two different chi-square distributions

Table A.5 provides some upper-tail percentiles for chi-square distributions.

	p = Area to Right of Chi-square Value								
df	*0.50*	*0.25*	*0.10*	*0.075*	*0.05*	*0.025*	*0.01*	*0.005*	*0.001*
1	0.45	1.32	2.71	3.17	3.84	5.02	6.63	7.88	10.83
2	1.39	2.77	4.61	5.18	5.99	7.38	9.21	10.60	13.82
3	2.37	4.11	6.25	6.90	7.81	9.35	11.34	12.84	16.27
4	3.36	5.39	7.78	8.50	9.49	11.14	13.28	14.86	18.47
5	4.35	6.63	9.24	10.01	11.07	12.83	15.09	16.75	20.51
6	5.35	7.84	10.64	11.47	12.59	14.45	16.81	18.55	22.46
7	6.35	9.04	12.02	12.88	14.07	16.01	18.48	20.28	24.32
8	7.34	10.22	13.36	14.27	15.51	17.53	20.09	21.95	26.12
9	8.34	11.39	14.68	15.63	16.92	19.02	21.67	23.59	27.88
10	9.34	12.55	15.99	16.97	18.31	20.48	23.21	25.19	29.59

From Utts, Jessica M. and Robert F. Heckard. Mind on Statistics, Fourth Edition. 2012. Used with permission.

Try It!

Consider the $\chi^2(4)$ distribution.

a. What is the mean for this distribution? __________

b. What is the median for this distribution? __________

c. How likely would it be to get a value of 4 or even larger?

d. How likely would it be to get a value of 10.3 or even larger?

The BIG IDEA

- The data consists of observed counts.
- We compute expected counts under the H_0 - these counts are what we would expect (on average) if the corresponding H_0 were true.
- Compare the observed and expected counts using the X^2 test statistic. The statistic will be a measure of how close the observed counts are to the expected counts under H_0. If this distance is large, we have support for the alternative H_a.

With this in mind, we turn to our first chi-square test of goodness of fit. We will derive the methodology for the test through an example. An overall summary of the test will be presented at the end.

Test of Goodness of Fit: Helps us assess if a particular discrete model is a good fitting model for a discrete characteristic, based on a random sample from the population.

Goodness of Fit Test

Scenario: We have one population of interest, say all cars exiting a toll road that has four booths at the exit.

Question: Are the four booths used equally often?

Data: One random sample of 100 cars, we record one variable *X,* which booth was used (1, 2, 3, 4). The table below summarizes the data in terms of the observed counts.

	Booth 1	**Booth 2**	**Booth 3**	**Booth 4**
Observed # cars	**26**	**20**	**28**	**26**

Note: This is only a one-way frequency table, not a two-way table as will be in the homogeneity and independence tests. We use the notation k = the number of categories or cells, here $k = 4$.

The null hypothesis:

Let p_i = (population) proportion of cars using booth i

H_0: p_1 = ________ , p_2 = ________ , p_3 = ________ , p_4 = ________ .

H_a: ____

The null hypothesis specifies a particular discrete model (mass function) by listing the proportions (or probabilities) for each of the k outcome categories.

The one-way table provides the OBSERVED counts. Our next step is to compute the EXPECTED counts, under the assumption that H_0 is true.

How to find the expected counts?

There were 100 cars in the sample and 4 booths.

If the booths are used equally often, H_0 is true, then we would expect

... _______ cars to use Booth #1

... _______ cars to use Booth #2

... _______ cars to use Booth #3

... _______ cars to use Booth #4

Expected Counts $= E_i = np_i$

Enter these expected counts in the parentheses in the table below.

	Observed Counts (Expected Counts)}			
	Booth 1	**Booth 2**	**Booth 3**	**Booth 4**
Number of cars	**26 ( )**	**20 ( )**	**28 ( )**	**26 ( )**

The X^2 test statistic

Next we need our test statistic, our measure of how close the observed counts are to what we expect under the null hypothesis.

$X^2 =$

Do you think a value of $X^2 =$ ___________ is large enough to reject H_0?

Let's find the *p*-value, the probability of getting an X^2 test statistic value as large or larger than the one we observed, assuming H_0 is true. To do this we need to know the distribution of the X^2 test statistic under the null hypothesis.

If H_0 is true, then X^2 has the χ^2 distribution with degrees of freedom =

.

***Find the p-value* for our tollbooth example:**

Observed X^2 test statistic value = __________ df = _______ .

Are the results statistically significant at the 5% significance level?
Conclusion at a 5% level: It appears that ....

Aside: Using our frame of reference for chi-square distributions.
Recall that if we have a chi-square distribution with df = degrees of freedom, then the mean is equal to df, and the standard deviation is equal to $\sqrt{2(df)}$
So, if H_0 were true, we would expect the X^2 test statistic to be about

give or take about _______________ .

Since we reject H_0 for large values of X^2, and we only observed a value of ,
even less than expected under H_0, we certainly do not have enough evidence to reject H_0.

Goodness of Fit Test Summary

Assume: We have 1 random sample of size n.
We measure one discrete response *X* that has k possible outcomes

Test: H_0: A specified discrete model for *X* → $p_1 = p_{10},\quad p_2 = p_{20},\quad \ldots,\quad p_k = p_{k0}$
H_a: The probabilities are not as specified in the null hypothesis.

Test Statistic: $$X^2 = \sum \frac{(\text{observed - expected})^2}{\text{expected}}$$
where expected $= E_i = np_{i0}$

If H_0 is true, then X^2 has a χ^2 distribution with $(k-1)$ degrees of freedom, where k is the number of categories. The necessary conditions are: at least 80% of the expected counts are greater than 5 and none are less than 1. Be aware of the sample size (pg 656).

Try It! Crossbreeding Peas

For a genetics experiment in the cross breeding of peas, Mendel obtained the following data in a sample from the second generation of seeds resulting from crossing yellow round peas and green wrinkled peas.

Yellow Round	Yellow Wrinkled	Green Round	Green Wrinkled
315	101	108	32

Do these data support the theory that these four types should occur with probabilities 9/16, 3/16, 3/16, and 1/16 respectively? Use $\alpha = 0.01$.

Try It! Desired Vacation Place

The AAA travel agency would like to assess if the distribution of *desired vacation place* has changed from the model of 3 years ago. A random sample of 928 adults were polled by the polling company *Ipsos* during this past mid-May. One question asked was "Name the one place you would want to go for vacation if you had the time and the money." The table displays the model for the distribution of desired vacation place 3 years ago and the observed results based on the recent poll.

	1 = Hawaii	2 = Europe	3 = Caribbean	4 = Other	Totals
Model 3 years ago	10%	40%	20%	30%	100%
Obs Counts from poll	124	390	125	289	928

a. Give the null hypothesis to test if there has been a significant change in the distribution of desired vacation place from 3 years ago.

b. The observed test statistic is ~ 31 and the *p*-value is less than 0.001. Interpret this *p*-value in terms of repeated random samples of 928 adults.

Test of Homogeneity: Helps us to assess if the distribution for one discrete (categorical) variable is the same for two or more populations.

Test of Homogeneity

Scenario: We have 2 populations of interest; preschool boys and preschool girls.
Question: Is Ice Cream Preference the same for boys and girls?

Data: 1 random sample of 75 preschool boys,
1 random sample of 75 preschool girls;
the two random samples are independent.

The table below summarizes the data in terms of the observed counts.
Observed Counts:

Ice Cream Preference	Boys	Girls
Vanilla (V)	**25**	**26**
Chocolate (C)	**30**	**23**
Strawberry (S)	**20**	**26**

Note: The column totals here were known in advance, even before the ice cream preferences were measured. This is a key idea for how to distinguish between the test of homogeneity and the test of independence.

The **null hypothesis**:

H_0: The distribution of ice cream preference is the **same** for the two populations, boys and girls.

A more mathematical way to write this null hypothesis is:

H_0: $P(X = i \mid population\ j) = P(X = i)$ for all i, j

where X is the categorical variable, in this case, ice cream preference.

As we can see, the null hypothesis is stating that the distribution of ice cream preference does not depend on (is independent of) the population we select from since the two distributions are the same.

The null hypothesis looks like: $P(A \mid B) = P(A)$, which is one definition of independent events, from our previous discussion of independence. This is why the test of homogeneity (comparing several populations) is really the same as the test of independence. The assumptions are different however.

For our homogeneity (comparing several populations) test, we assume we have independent random samples, one from each population, and we measure 1 discrete (categorical) response. For the independence test (discussed later) we will assume we have just 1 random sample from 1 population, but we measure 2 discrete (categorical) responses.

Getting back to **ICE CREAM** ... The table provides the OBSERVED counts. Our next step is to compute the EXPECTED counts, under the assumption that H_0 is true.

How to find the expected counts?

Let's look at those who preferred Strawberry first.

Strawberry: Since there were ______ children who preferred *Strawberry* overall, if the distributions for boys and girls are the same (H_0 is true),

then we would expect _________ of these children to be boys

and the remaining ___________ of these children to be girls.

Note that our sample sizes were the same, 75 boys and 75 girls, 50% of each. If they were not 50-50, we would have to adjust the expected counts accordingly. Let's do the same for the Vanilla and Chocolate preferences.

Chocolate: Since there were ______ children who preferred Chocolate overall, if the distributions for boys and girls are the same (H_0 is true), then we would expect _________ of these children to be boys

and the remaining ___________ of these children to be girls.

Vanilla: Since there were _________ children who preferred Vanilla overall, if the distributions for boys and girls are the same (H_0 is true), then we would expect _________ of these children to be boys

and the remaining ___________ of these children to be girls.

Enter these expected counts in the parentheses in the table below.

Observed Counts (Expected Counts)

Ice Cream Preference	**Boys**	**Girls**	Total
Vanilla (V)	**25( )**	**26 ( )**	51
Chocolate (C)	**30( )**	**23 ( )**	53
Strawberry (S)	**20( )**	**26 ( )**	46
Total	75	75	150

A Closer Look at the Expected Counts:

Let's look at how we actually computed an expected count so we can develop a general rule: If H_0 were true (i.e., no difference in preferences for boys versus girls), then our *best* estimate of the P(a child prefers vanilla) = 51/150. Since we had 75 boys, under no difference in preference, we would expect 75 x (51/150) to prefer vanilla. That is, the expected number of boys preferring vanilla = $\frac{(75)(51)}{150} = \frac{\text{(row total)(column total)}}{\text{Total n}}$. This quick recipe for computing the expected counts under the null hypothesis is called the **Cross-Product Rule**.

The X^2 test statistic

Next we need to compute our test statistic, our measure of how close the observed counts are to what we expect under the null hypothesis.

$$X^2 = \frac{(25-25.5)^2}{25.5} + \dots$$

The larger the test statistic, the "bigger" the differences between what we observed and what we would expect to see if H_0 were true. So the larger the test statistic, the more evidence we have against the null hypothesis.

Is a value of $X^2 =$ __________ large enough to reject H_0?

We need to find the *p*-value, the probability of getting an X^2 test statistic value as large or larger than the one we observed, assuming H_0 is true. To do this we need to know the distribution of the X^2 test statistic under the null hypothesis.

If H_0 is true, then X^2 has the χ^2 distribution with degrees of freedom =

Brief motivation for the degrees of freedom formula:

***Find the p-value* for our ice cream example:**

Observed X^2 test statistic value = ________________ df = __________________

Decision at a 5% significance level: (circle one) **Reject H_0** **Fail to reject H_0**

Conclusion: It appears that ....

Test of Homogeneity Summary (Comparison of Several Populations)

Assume: We have C independent random samples of size $n_1, n_2, ..., n_c$ from C populations.
We measure 1 discrete response X that has r possible outcomes.

Test:
H_0: The distribution for the response variable X is the same for all populations.

Test Statistic: $X^2 = \sum \frac{(\text{observed - expected})^2}{\text{expected}}$

where expected $= \frac{(\text{row total})(\text{column total})}{\text{Total } n}$

If H_0 is true, then X^2 has a χ^2 distribution with $(r-1)(c-1)$ degrees of freedom. The necessary conditions are: at least 80% of the expected counts are greater than 5 and none are less than 1.

Try It! What is your Decision?

For a chi-square test of homogeneity, there are 3 populations and 4 possible values of the discrete characteristic.

If H_0 is true, that is, the distribution for the response is the same for all 3 populations, what is the expected value of the test statistic?

Try It! Treatment for Shingles

An article had the headline "For adults, chicken pox vaccine may stop shingles". A clinical trial was conducted in which 420 subjects were randomly assigned to receive the chicken pox vaccine or a placebo vaccine. Some side effects of interest were swelling and rash around the injection site. Consider the following results for the swelling side effect.

Treatment Group * Swelling Status Crosstabulation

Count

		Swelling Status			Total
		major swelling	minor swelling	no swelling	
Treatment Group	vaccine	54	42	134	230
	placebo	16	32	142	190
Total		70	74	276	420

Chi-Square Tests

	Value	df	Asymp. Sig. (2-sided)
Pearson Chi-Square	18.571	2	.00009
Likelihood Ratio	19.556	2	.00006
Linear-by-Linear Association	17.696	1	.00003
N of Valid Cases	420		

a. Give the name of the test to be used for assessing if the distribution of swelling status is the same for the two treatment populations.

b. Based on the above data, among those chicken pox vaccinated subjects, what percent had major swelling around the injection site?

c. Based on the above data, among those placebo vaccinated subjects, what percent had major swelling around the injection site?

d. Assuming the distribution of swelling status is the same for the two treatment populations, how many chicken pox vaccinated subjects would you expect to have major swelling around the injection site? ***Show your work.***

e. Use a level of 0.05 to assess if the distribution of swelling status is the same for the two treatment populations.
Test Statistic Value: ___________ *p*-value: _____________
Thus, the distribution of swelling status (circle your answer): **does** **does not** appears to be the same for the two treatment populations.

Test of Independence: Helps us to assess if two discrete (categorical) variables are independent for a population, or if there is an association between the two variables.

Test of Independence

Scenario: We have one population of interest - say factory workers.

Question: Is there a relationship between smoking habits and whether or not a factory worker experiences hypertension?

Data: 1 random sample of 180 factory workers, we measure the two variables:

X = hypertension status (yes or no)

Y = smoking habit (non, moderate, heavy)

The table below summarizes the data in terms of the observed counts.

Observed Counts:

		Y= Smoking Habit		
X=		Non	Mod	Heavy
Hyper	Yes	21	36	30
Status	No	48	26	19

Get the row and column totals.

Note: neither row nor column totals were known in advance before measuring hypertension and smoking habit. We only know the overall total of 180.

The **null hypothesis**:

H_0: There is no association between smoking habit and hypertension status for the population of factory workers.
(or The two factors, smoking habit and hypertension status, are independent for the population.)

One more mathematical way to write this null hypothesis is:

$$H_0: P(X = i \text{ and } Y = j) = P(X = i)P(Y = j)$$

The null hypothesis looks like: $P(A \text{ and } B) = P(A)P(B)$, which is one definition of independent events, from our previous discussion of independence.

Getting back to our FACTORY WORKERS ...

The two-way table provides the OBSERVED counts. Our next step is to compute the EXPECTED counts, under the assumption that H_0 is true. The expected counts and the test statistic are found the same way as for the homogeneity test.

Cross-Product Rule: **Expected Counts** $= \dfrac{\text{(row total)(column total)}}{\text{Total } n}$

Compute and enter these expected counts in the parentheses in the table below.

Observed Counts (Expected Counts):

		Y= Smoking Habit			
X=		Non	Mod	Heavy	
Hyper Status	Yes	21 ( )	36 ( )	30 ( )	87
	No	48 ( )	26 ( )	19 ( )	93
		69	62	49	180

The X^2 test statistic

Our measure of how close the observed counts are to what we expect under the null hypothesis.

$$X^2 = \frac{(21-33.35)^2}{33.35} + \ldots$$

Do you think a value of $X^2 =$ ____________ is large enough to reject H_0?

The next step is to find the *p*-value, the probability of getting an X^2 test statistic value as large or larger than the one we observed, assuming H_0 is true. To do this we need to know the distribution of the X^2 test statistic under the null hypothesis.

If H_0 is true, then X^2 has the χ^2 distribution with degrees of freedom =_______

Aside: Using our frame of reference for chi-square distributions.

If H_0 were true, we would expect the X^2 test statistic to be about _________ give or take about ______________ .

About how many standard deviations is the observed X^2 value of 14.5 from the expected value under H_0? What do you think the decision will be?

Find the *p*-value for our factory worker example:

Observed X^2 test statistic value = ____________ df = ____________

Find the *p*-value and use it to determine if the results are statistically significant at the 1% significance level.

Conclusion at a 1% level: It appears that ....

Test of Independence Summary

Assume: We have 1 random sample of size n.
We measure 2 discrete responses:
X which has r possible outcomes
and *Y* which has c possible outcomes.

Test: H_0: The two variables *X* and *Y* are independent for the population.

Test Statistic: $X^2 = \sum \frac{(\text{observed - expected})^2}{\text{expected}}$

where expected $= \frac{(\text{row total})(\text{column total})}{\text{Total } n}$

If H_0 is true, then X^2 has a χ^2 distribution with $(r-1)(c-1)$ degrees of freedom. The necessary conditions are: at least 80% of the expected counts are greater than 5 and none are less than 1.

Relationship between Age Group and Appearance Satisfaction

Are you satisfied with your overall appearance? A random sample of 150 women were surveyed. Their answer to this question (Yes or No) was recorded along with their age category (1 = under 30, 2 = 30 to 50, and 3 = over 50). SPSS was used to generate the following output from the data.

Are You Satisfied? * Age Group Crosstabulation

Count

		Age Group			Total
		under 30	30 to 50	over 50	
Are You Satisfied?	Yes	38	30	34	102
	No	10	29	9	48
Total		48	59	43	150

Chi-Square Tests

	Value	df	Asymp. Sig.
Pearson Chi-Square	13.149	2	.001
Likelihood Ratio	13.039	2	.001
Linear-by-Linear Association	.018	1	.893
N of Valid Cases	150		

a. Give the name of the test to be used for assessing if there is a relationship between age group and appearance satisfaction.

b. Assuming there is no relationship between age group and appearance satisfaction, how many old women (over 50) would you expect to be satisfied with their appearance? ***Show your work.***

c. Assuming there is no relationship between age group and appearance satisfaction, what is the expected value of the test statistic?

d. Use a level of 0.05 to assess if there is a significant relationship between age group and appearance satisfaction.

Test Statistic Value: _______________ *p*-value: __________________

Thus, there (circle your answer): **does** **does not**

appear to be an association between age group and appearance satisfaction.

2x2 Tables – a special case of the two proportion z test

Section 15.2 (pages 593-600 of the textbook) discuss some of the special circumstances that apply to analyzing 2x2 tables.

- The z-test for comparing two population proportions is the same as the chi-square test provided the alternative is two-sided. The z-test would need to be performed for one-sided alternatives.
- When the conditions for the z-test or chi-square test are not met (sample sizes too small) there is another alternative test called the Fisher's Exact Test.

Stat 250 Formula Card:

Chi-Square Tests	
Test of Independence & Test of Homogeneity	**Test for Goodness of Fit**
Expected Count $E = \text{expected} = \frac{\text{row total} \times \text{column total}}{\text{total } n}$	**Expected Count** $E_i = \text{expected} = np_{i0}$
Test Statistic $X^2 = \sum \frac{(O-E)^2}{E} = \sum \frac{(\text{observed} - \text{expected})^2}{\text{expected}}$ df = $(r-1)(c-1)$	**Test Statistic** $X^2 = \sum \frac{(O-E)^2}{E} = \sum \frac{(\text{observed} - \text{expected})^2}{\text{expected}}$ df = $k-1$
If Y follows a $\chi^2(df)$ distribution, then E(Y) = df and Var(Y) = 2(df).	

Additional Notes

A place to ... jot down questions you may have and ask during office hours, take a few extra notes, write out an extra practice problem or summary completed in lecture, create your own short summary about this chapter.

Made in United States
Orlando, FL
04 November 2026

90218658R00140